DISCARDED

# MRI Techniques

# MRI Techniques

Vincent Perrin

*Series Editor*
*Henri Maître*

WILEY

First published 2013 in Great Britain and the United States by ISTE Ltd and John Wiley & Sons, Inc.

ISTE Ltd
27-37 St George's Road
London SW19 4EU
UK

www.iste.co.uk

John Wiley & Sons, Inc.
111 River Street
Hoboken, NJ 07030
USA

www.wiley.com

Library of Congress Control Number: 2013946298

British Library Cataloguing-in-Publication Data
A CIP record for this book is available from the British Library
ISBN: 978-1-84821-503-0

Printed and bound in Great Britain by CPI Group (UK) Ltd., Croydon, Surrey CR0 4YY

# Table of Contents

# Chapter 1

# Flow

## 1.1. Blood

### 1.1.1. *Characteristics of bloodflow*

Bloodflow is characterized by four parameters:

– its velocity;

– its acceleration;

– its direction;

– its orientation.

Arterial flow has the highest velocity.

This velocity varies with the phases of the cardiac cycle: it may be very high (150-175 cm/s during systole, the period of contraction of the myocardium) but also almost null (at the end of diastole, the period of relaxation of the myocardium).

The flow in the cerebral arteries is much slower: 40-70 cm/s.

Finally, venous flow is the slowest: generally less than 20 cm/s.

Bloodflow runs from the organs to the heart in venous circulation and from the heart to the organs in arterial circulation.

### 1.1.2. *Laminar flow and turbulent flow*

We can distinguish two main types of flow:

– A laminar flow is found when the velocities are relatively low. It is characterized by a distribution of velocities which are all parallel and whose profile is parabolic: the velocity of the fluid is maximum at the center and almost null when against the walls.

**Figure 1.1.** *Profile of velocities in a laminar flow*

– A turbulent flow is a chaotic form of flow. The velocity then exhibits a vortical nature: local circular motions arise.

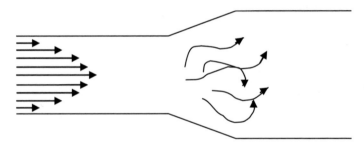

**Figure 1.2.** *Profile of velocities in a turbulent flow*

There are many reasons why a laminar flow could become a turbulent flow:

– the velocity of the flow surpasses a critical value;

– the structure of the vessels changes (see Figure 1.2).

Venous flow is laminar with a velocity that is constant overall.

Arterial flow is an intermediary case: it is laminar in diastole and turbulent during systole.

A turbulent flow causes significant artifacts in images of the flow, which we shall now go on to discuss in detail.

## 1.2. Basic phenomena in angiography

The flow phenomena which we are going to describe in this section enable us to spontaneously view blood vessels (without the injection of a contrast-enhancing product); the effect they have on the image depends on the sequence used (spin echo (SE) or gradient echo (GE)), on the parameters of that sequence ($T_R$, $T_E$...) but also on the particular parameters of the flow itself: its velocity, the orientation of the vessel in relation to the slice, etc.

Thus, we can construct a form of imaging relating to the flow: magnetic resonance angiography, or MRA.

### 1.2.1. *Time of Flight (TOF)*

When the vessel runs through the slice, the intensity of the flow signal depends on the time-of-flight of the protons $T_t$, i.e. the time taken to traverse the thickness $\Delta z$ of the plane being imaged at velocity V.

#### 1.2.1.1. *Phenomenon of flow void in a spin-echo sequence*

We work on a *spin-echo* sequence where the 90° and 180° pulses are selective in the given slice.

*With stationary protons:* they experience both pulses and are therefore able to generate a signal.

*With moving protons* (in the bloodstream), two scenarios may arise:

– Either they remain in the slice ($T_t < T_E/2$) and are therefore struck by both pulses to generate a signal. For a vessel perpendicular to the slice, these protons have a velocity of $V < \Delta z/(T_E/2)$: this is qualified as a "slow" flow.

– Or they leave the slice entirely before the transmission of the 180° pulse (i.e. $T_t < T_E/2$). They are then replaced within the slice by protons which were not subjected to the initial 90° pulse and which therefore do not generate a signal. (The extreme case, $T_t = T_E/2$, is represented in Figure 1.3). For a vessel perpendicular to the slice, these protons have a velocity of $V > \Delta z/(T_E/2)$: this is qualified as a "fast" flow.

NOTE.– "Fast flow" is not necessarily synonymous with "arterial flow": flow in large veins is also included in this category.

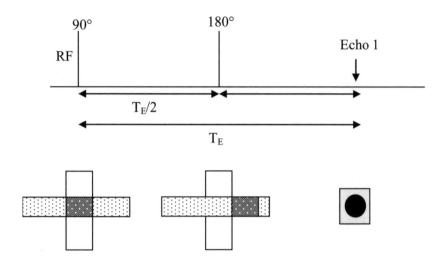

**Figure 1.3.** *Case of $T_t = T_E/2$: the fate of protons experiencing the 90° excitation pulse (in white dots) and cross-sectional image of the flow*

Flow void is the most commonly encountered phenomenon in flow imaging: this phenomenon accounts for the fact that many blood vessels are spontaneously visible on an MRI image by the absence of signal (by the "wall / light" contrast).

*1.2.1.2. Phenomenon of flow-related (or paradoxical) enhancement*

1.2.1.2.1 The phenomenon

We work on a spin-echo or gradient-echo sequence.

We suppose the $T_R$ to be short: for this reason, there is repetition within a sufficiently short time period of the 90° excitation pulses.

*With stationary protons*: they have already been excited by the 90° pulse from the previous cycle (i.e. from the previous $T_R$ interval): therefore, their longitudinal magnetization is low and they generate little signal.

*With moving protons*: if their velocity is sufficient, "fresh" protons, which have not been subjected to the 90° pulse from the previous cycle, enter into the slice: their longitudinal magnetization vector is maximum and the signal engendered by these protons is therefore maximum as well: this is the *phenomenon of paradoxical enhancement*.

## 1.2.1.2.2. The $T_R$ dictates this phenomenon

The phenomenon of paradoxical enhancement is all the more marked when the TR is short: *gradient-echo sequences* (see below) are therefore more sensitive to this phenomenon, and therefore we shall limit ourselves to this type of sequence in our discussion below.

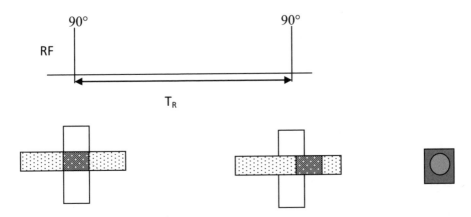

**Figure 1.4.** *Case of $T_t > T_R$: fate of protons in the flow that experience the 90° excitation pulse (in white dots) and cross-sectional image of the flow. The vertical slab indicates all of the protons in the flow (in black dots) or outside of it (in white) affected by the different pulses*

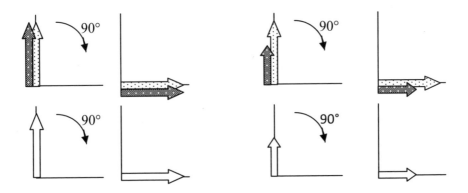

**Figure 1.5.** *Case of $T_t > T_R$: longitudinal and transversal magnetization of the protons in the flow (top) and outside of it (bottom)*

This phenomenon is maximum when all the protons excited during a cycle leave the slice and are replaced by "fresh" protons in the next cycle ($T_t < T_R$).

For a vessel running perpendicular to the slice, these protons have a velocity of $V > \Delta z/T_R$. (The extreme case, with maximum contrast: $T_t = T_R$, is represented in Figure 1.6).

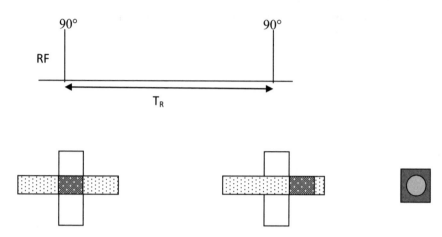

**Figure 1.6.** *Case of $T_t = T_R$: fate of the protons in the flow that experience the 90° excitation pulse (in white dots) and cross-sectional image of the flow. The vertical strip indicates all the protons in the flow (in black dots) or outside of it (in white) affected by the different pulses*

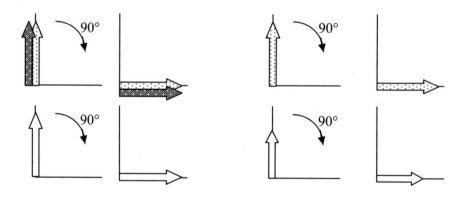

**Figure 1.7.** *Case of $T_t = T_R$: longitudinal and transversal magnetization of the protons in the flow (top) and outside of it (bottom)*

NOTE.– For *spin-echo sequences*, if $T_R$ = 1s and $\Delta z$ = 1 cm, then paradoxical enhancement is maximum for velocities $V > \Delta z/T_R$, i.e. $V > 1$ cm/s (this velocity, which is very slow, corresponds to venous flow at the end of diastole). The contrast obtained (the difference between flows of different velocities) is therefore very poor – all the more so if the stationary protons in the tissue are not saturated with a similarly long $T_R$.

This indeed reaffirms the necessity of working in a gradient-echo regime if we wish to use the paradoxical signal.

1.2.1.2.3. Single slice and multi-slice

In a multi-slice sequence, the paradoxical signal predominates over the first slice(s) (or the last) depending on the orientation of the flow (i.e. on the "flow entry slices"): such is the case for slice A in the diagram below.

Indeed, in planes which are a long way from the flow entry slices, the protons have been saturated by previous 90° excitations in the upstream slices (much like stationary protons): such is the case for slices C, D and E.

However, the paradoxical signal may also be visible on the *first* intermediary slices (slice B): sometimes we note the remnant presence of an intravascular "halo" signal in these slices. This halo is due to the differences in velocity of the protons at the center of the vessel (fast flow) and at the edges of the vessel (slow flow) in the case of a laminar flow with parabolic profile.

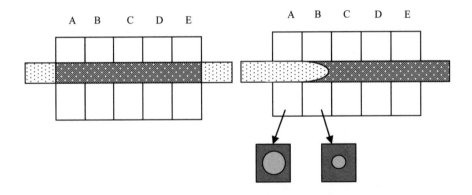

**Figure 1.8.** *Paradoxical enhancement in a multi-slice sequence*

NOTE.– The gradient-echo sequence, which is favored when we wish to use the paradoxical signal, usually enables us to take very short $T_R$ values, which generally lends itself to single-slice processing: thus, each successive slice behaves like a new "entry slice".

### 1.2.2. *Phenomenon of dephasing of circular spins*

1.2.2.1. Dephasing of protons in a unipolar gradient

We consider protons whose velocity v is supposed to be constant (i.e. with zero acceleration) and direction $x$. The phase of the protons will depend only on the gradient in that direction: $G_x$.

At a time $t_e$ taken as a reference point, the position of the protons is $x(t_e)$ and their velocity in the direction of $G_x$ is $v(t_e)$.

We wish to discover the phase $\phi(t_m)$ of the protons at a time $t_m > t_e$.

Thus, let us take a gradient of time-length $\tau$ and constant amplitude $A$.

The gradient is switched on at time $T - \dfrac{\tau}{2}$ and switched off at $T + \dfrac{\tau}{2}$.

Thus $\varphi(t_m) = \displaystyle\int_{T-\frac{\tau}{2}}^{T+\frac{\tau}{2}} \gamma \cdot G_x \cdot x(t)dt = \int_{T-\frac{\tau}{2}}^{T+\frac{\tau}{2}} \gamma \cdot G_x \cdot [x(t_e) + v \cdot (t - t_e)]dt$ ;

Therefore: $\varphi(t_m) = \gamma \cdot A \cdot \left( x(t_e) \cdot \tau + \left[ \dfrac{v \cdot (t-t_e)^2}{2} \right]_{T-\frac{\tau}{2}}^{T+\frac{\tau}{2}} \right) = \gamma \cdot A \cdot (x(t_e) \cdot \tau + v \cdot (T - t_e) \cdot \tau)$

However, $x(T) = x(t_e) + v \cdot (T - t_e)$, so $\boxed{\varphi(t_m) = \gamma \cdot A \cdot x(T)}$

NOTE.– The phase is independent of $t_m$.

1.2.2.2. *Dephasing of the protons in a bipolar gradient*

Based on this simple gradient, we can construct gradients which behave differently in relation to stationary or moving protons.

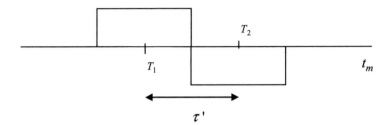

**Figure 1.9.** *Bipolar gradient*

The amplitudes of the two lobes of the bipolar gradient shown in Figure 1.9 are respectively $A$ and $-A$.

The phase obtained at time $t_m$ is thus given by:

$$\varphi(t_m) = \gamma \cdot A \cdot x(T_1) - \gamma \cdot A \cdot x(T_2) = -\gamma \cdot A \cdot v \cdot (T_2 - T_1)$$

Hence: $\boxed{\varphi(t_m) = -\gamma \cdot v \cdot A \cdot \tau'}$

For stationary protons: $\varphi = 0$, the protons are refocused.

(NOTE.– we quite deliberately discount the dephasing due to the inhomogeneities of the field $\varphi_B$ and to the Foucault currents $\varphi_e$.)

For moving protons, their movement in the direction of the gradient makes it impossible to re-phase their dephasing: their phase is proportional to their velocity.

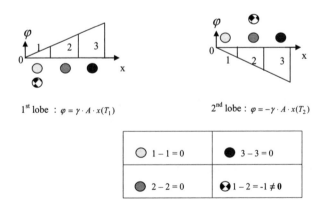

**Figure 1.10.** *Phases acquired by the different protons with the application of each lobe, and total phases obtained (table at bottom). The stationary protons are illustrated in grayscale; moving ones are illustrated in black and white*

NOTE.– it is not necessary for the two lobes of the gradient to be adjacent: the time $\tau'$ is, in this case, the length of time between the centers of gravity of each of the two lobes, and the formula $\phi(t_m) = -\gamma \cdot v \cdot A \cdot \tau'$ remains valid.

### 1.2.2.3. Form of velocity-insensitive gradients: flow compensation gradient

The frequency gradients (selection of slice and readout) mark protons when they are applied. Generally, they are bipolar in form, to facilitate the protons' readjustment phase.

However, this rephasing is possible only for stationary protons, rather than moving protons, whose dephasing is due to their velocity, as shown by section 1.2.2.2.

In order to compensate for the dephasing due to the displacement of protons circulating at constant velocity, we need to *add* lobes to the frequency gradients.

In order to do this, the bipolar gradient is altered to possess three lobes in the ratio 1:2:1.

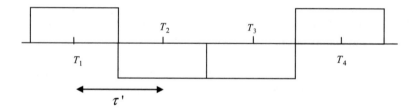

**Figure 1.11.** *Alteration of the bipolar gradient to compensate for the phase shifting due to the displacement of protons and thereby obtain a flow compensation gradient*

This type of gradient is indeed the sum of two bipolar gradients; the first creating a phase-shift $\phi_1(t_m) = -\gamma \cdot v \cdot A \cdot \tau'$ of the moving protons, and the second an opposite phase-shift: $\phi_2(t_m) = +\gamma \cdot v \cdot A \cdot \tau'$. Finally, $\phi_{TOTAL}(t_m) = 0$: there is rephasing of the stationary protons *and* the moving protons.

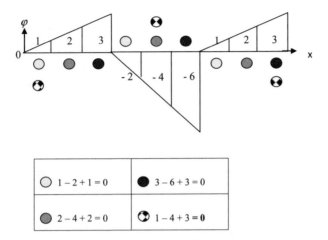

**Figure 1.12.** *Phases acquired by the different protons when a flow compensation gradient is applied*

NOTE.– Obviously, it is also possible to compensate for the dephasing caused by constant acceleration. We need only add extra lobes.

However, the disadvantage of adding lobes to the reading gradient is that it prolongs the $T_E$, which also entails an increase in the dephasing of the spins (a decrease in $T_2$*!) and therefore weakens the signal.

For this reason, only phase-shifts linked to a constant velocity tend to be corrected, with $T_E$ being kept as short as possible.

## 1.3. Artifacts relating to the flow

The artifacts described in this section are those which can be "conventionally" encountered, without necessarily intending to perform angiography. (Thus, the artifacts relating to MRA sequences, such as aliasing, are discussed later on in the section relating to the sequence itself).

Here, therefore, we are interested in *"conventional" artifacts of flow.*

### 1.3.1. *Artifact of pulsatile flow of blood or cerebrospinal fluid (CSF)*

1.3.1.1. *Causes...*

The varying velocity of CSF or blood over the course of the cardiac cycle means that the intensity of the signal measured at the same voxel will be *periodic* over time:

– the renewal of non-saturated protons is surely to be found during the period of systole (high average flow velocity) and absent during diastole (low flow velocity);

– the dephasing of spins in the same voxel may be great during systole (high average flow velocity, so high velocity contrast and therefore phase contrast) and low during diastole (low flow velocity, so low velocity contrast and therefore phase contrast).

### 1.3.1.2. ...and consequences

Thus, the pulsatile flow of blood or CSF causes artifacts which often take the form of *ghost images* in the direction of phase coding, the intensity of which is variable depending on the extent of the effects of TOF or dephasing of the moving protons.

It is possible to see these artifacts in the case of a spin-echo sequence: during systole, the moving protons excited by the 90° pulse exit the slice and are therefore not touched by the 180° pulse; during diastole, however, the signal is strong because the average velocity of the protons is low.

Nevertheless, these artifacts are often more pronounced on gradient-echo sequences, where because of paradoxical enhancement, the intensity of the signal may be very great.

We regularly find this type of artifact in transverse slices containing the aorta, as shown in Figure 1.13.

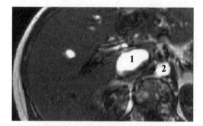

**Figure 1.13.** *Axial view of the abdomen. Flow artifact in the direction of the phase coding gradient due to the paradoxical enhancement of the signal from the vena cava (1) and the aorta (2)*

It is important to reduce the impact of this artifact, as a diagnosis (such as the detection of a vertebral angioma, for instance) could be made difficult if ghost images are projected onto the image of interest.

### 1.3.1.3. *Solutions*

There are various ways to decrease ghost images:

– *ECG synchronization* can be employed to always trigger the acquisitions at the same moment in the cardiac cycle, which reduces signal modulation due to a difference in the average velocity of the bloodflow in the vessels.

– *Flow compensation gradients* minimize pulsatile artifacts in the blood or CSF by reducing the dephasing due to the fairly rapid displacements of the protons within the same voxel.

– *Pre-saturation* involves applying saturation slabs to the slices upstream of the one under examination.

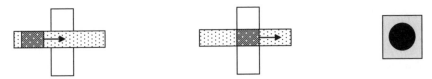

**Figure 1.14.** *Pre-saturation strip (white) and its effect on the flow signal*

The role of these slabs is to saturate the blood in the vessels before it arrives at the slice of interest, so that the moving protons do not emit a signal (hence there can be no modulation of the intensity of that signal).

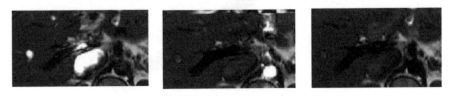

**Figure 1.15.** *Elimination of the artifact of flow encountered previously (see Figure 1.14), using pre-saturation slabs applied...above the slice: elimination of the ghost image of the aorta but not of the vena cava (top left)... below the slice: elimination of the ghost image of the vena cava but not of the aorta (top right)... above and below the slice: elimination of both artifacts (bottom)*

NOTE.– Pre-saturation also enables us to select a flow on the basis of its orientation: for instance, we could eliminate venous flow from consideration, keeping only arterial flow.

– Finally, we can *invert the gradients of phase coding and frequency*: this does not prevent ghosting from happening, but moves the ghost image to a different area in the image of interest. Thereby, we are able to be sure that it is indeed an artifact rather than a diseased structure.

### 1.3.2. *Fluid location error*

#### 1.3.2.1. *Description*

In an SE or GE sequence, the encoding in the direction of the phase gradient is done before that of the frequency gradient. This causes an error in locating the moving protons, as we shall shortly see:

– Phase coding takes place after a time $T$ after the excitation pulse; we shall consider the moment when the row $k_y = n$ is acquired. The increment of the gradient $G_y$ and its duration are respectively notated as $\Delta G_y$ and $\tau$. The phase of the protons which at that time are in position $(x(T), y(T))$ is therefore:

$$\varphi = \gamma \cdot n \cdot \Delta G_y \cdot y(T) \cdot \tau$$

– At the time of the gradient echo $G_x$, i.e. at time $T_E$, the protons are displaced and are now *located* in position $[x(T_E), y(T_E)]$, but are *recorded* in position

$$[x(T_E), y(T)] = [x(T_E), y(T_E) - v_y \cdot (T_E - T)] \qquad [1.1]$$

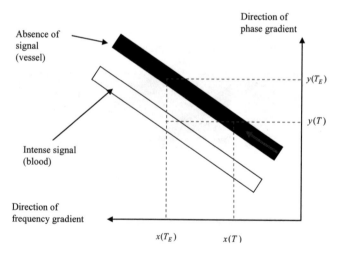

**Figure 1.16.** *Erroneous recording of the position of the vessel and appearance of a row with no signal*

The signal is therefore shifted in the direction of the phase gradient: we then see the appearance of a row with no signal, which is the vessel, bounded by a row with intense signal: that of the protons of the blood.

1.3.2.2. *Parameters influencing the artifact of location*

1.3.2.2.1. Orientation of the flow

For most SE or FE sequences used, it is possible to determine the orientation of blood circulation when an artifact of location is shown.

Indeed, we can see that the signal row is positioned alongside the signal-free row given by the component of the flow in the direction of the frequency gradient (indicated by the dotted arrow in the diagrams below).

This artifact of location is therefore widely used in clinical practice, particularly in cases of vascular malformation.

1.3.2.2.2. Echo time

The error in location of the flow is greater for vessels situated within the slice being examined. It is accentuated when the echo time $T_E$ is increased, as we can see from equation [1.1].

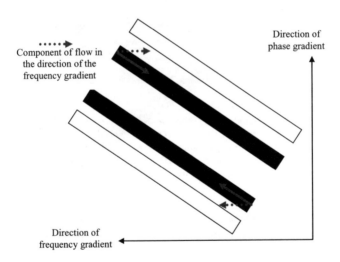

**Figure 1.17.** *Position of the artifact of location and orientation of flow*

1.3.2.3. *Clinical reality*

As is indicated by equation [1.1], the blood signal shifts proportionally to the component of velocity of the flow along the phase gradient axis: $v_y$ .

As vessels are not linear, this component varies with the position of the fluid: the artifact of shift is therefore also dependent on the position of the fluid (for instance, it is non-existent if $v_y = 0$); hence, there is more than a simple displacement: there is *distortion of the blood vessels* subjected to this artifact.

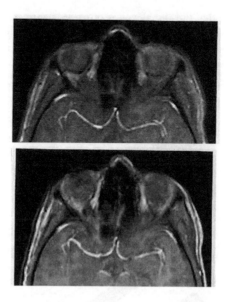

**Figure 1.18.** *Distortion in the case of a gradient-echo sequence showing the anterior cerebral artery. These images are obtained with identical echo times, but different directions for the phase gradients (left to right (LR) in the first image above and anterior to posterior (AP, front to back) in the second). (From [VLA 03]. Reproduced with kind permission from Springer)*

If we compare the images in Figure 1.18, obtained with a gradient-echo sequence, we note that the incurvate form of the anterior cerebral artery varies.

In the upper image, the distance between the two branches of the incurvate part is greater: this results from a shift of these two branches in the direction of the phase gradient (LR direction) and in opposite orientations.

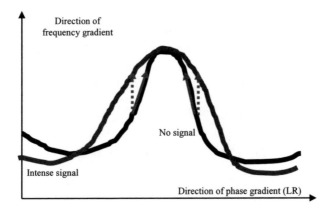

**Figure 1.19.** *Distortion in the case of an LR phase gradient.*
*The orientation of the flow is shown by the velocity vectors in solid line.*
*The dotted vectors indicate the displacement of the signal*

In the image below, the displacements of these two branches are in the AP direction: the two branches are shifted downwards (this shift is greater when the component $v_y$ is large, i.e. the bifurcation peak: this gives the peak a squashed appearance).

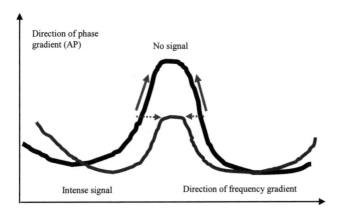

**Figure 1.20.** *Distortion in the case of an AR phase gradient.*
*Notations identical to those in Figure 1.19*

1.3.2.4. *Solution*

To correct the artifact of location is to correct the phase given by the application of the gradient $G_y$, as though the gradient had in reality been measured at time $T_E$ .

– Without correction, at time $T$ , the phase is: $\phi(T) = \gamma \cdot n \cdot \Delta G_y \cdot y(T) \cdot \tau$

– At time $T_E$ , the phase is:

$$\phi(T_E) = \gamma \cdot n \cdot \Delta G_y \cdot y(T_E) \cdot \tau = \gamma \cdot n \cdot \Delta G_y \cdot \left( y(T) + v_y \cdot (T_E - T) \right) \cdot \tau$$

Therefore,

$$\phi(T_E) = \phi(T) + \gamma \cdot n \cdot \Delta G_y \cdot v_y \cdot (T_E - T) \cdot \tau$$

In order for it to seem to have been measured at time $T_E$ , we should have $\phi(T) = \phi(T_E)$ .

Thus, to $\varphi(T)$ we need to add the correction $\Delta\phi = \gamma \cdot v_y \cdot n \cdot \Delta G_y \cdot (T_E - T) \cdot \tau$

However, we know (see section 1.2.2.2) that in the case of a bipolar gradient of duration $\tau$ and applied in direction $y$, the phase obtained is $-\gamma \cdot v_y \cdot A \cdot \tau$.

*Thus, during the coding of the row $k_y = n$, we need to add to the "normal" phase gradient a bipolar gradient of duration $\tau$ and amplitude $-n \cdot \Delta G_y \cdot (T_E - T)$.*

We note that it is thus possible to correct the artifact of location for a flow with unknown velocity (as the above formula makes no reference to that velocity).

NOTE.– This phase correction is costly in terms of time. Obviously, if a relatively short echo time is chosen, this correction cannot be done.

### 1.3.3. *Other artifacts*

1.3.3.1. *Artifact due to a velocity gradient*

1.3.3.1.1. Causes and consequences

On the walls of the vessels, there is a significant velocity gradient (see 1.1.2); as indicated in section 1.3.1 (ghost images due to pulsatile flow), this causes significant dephasing within the same voxel, and therefore a loss of signal on the periphery of the vessel. Its diameter therefore appears reduced; this artifact manifests itself mainly on axial slices.

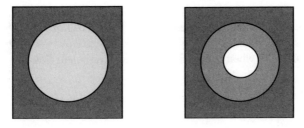

**Figure 1.21.** *Vessel without (left) and with (right) a velocity gradient*

This intravoxel phase dispersion is also observed when there is turbulence or acceleration of the flow (e.g. in cases of stenosis) or in the vicinity of a magnetic field gradient.

1.3.3.1.2. Solution

A flow compensation gradient can considerably reduce this artifact (the same way it reduces the artifact due to pulsatile flow).

1.3.3.2. *Another artifact encountered in an SE sequence*

With an SE sequence with multiple (and symmetrical) echoes, a special case is represented by the rephasing of the protons on even-numbered echoes.

Indeed, the dephasing of the protons, due to the flow and observable with the first echo, is exactly compensated by the second 180° pulse. Thus, the signal is attenuated in the first echo, but in the second echo (and all symmetrical even-numbered echoes), we shall see "recuperation" of the signal.

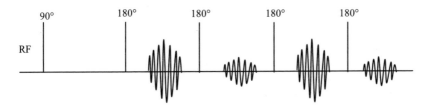

**Figure 1.22.** *Intensity of the different echoes obtained in the presence of a flow*

## 1.4. MRA sequences

The physical phenomena described in section 1.2 form the basis of the technique of MRA.

Depending on the underlying phenomenon, these techniques will be more or less sensitive to certain velocities of flow, or to certain types of flow. In particular, they will run into difficulty when faced with complex or turbulent flows.

### 1.4.1. *Time of Flight (TOF)*

1.4.1.1. *Technique*

1.4.1.1.1. Principle

In time-of-flight MRA, we use *gradient-echo* sequences to favor the signal from the flows by saturating the signal from stationary tissues with very short TR times: thereby, the longitudinal magnetization of these tissues does not have time to recover, and the signal generated by them fades, encouraging the phenomenon of slice entry: as the circulating blood entering into the slice being examined has not been saturated, its longitudinal magnetization is maximum. The signal from the bloodflow is therefore stronger than that from the stationary tissues: this is known as "white blood" imaging (see section 1.2.1.2).

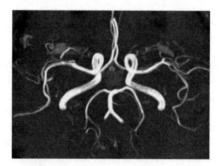

**Figure 1.23.** *Circle of Willis in white-blood TOF imaging (taken from [IRM])*

It is also possible to use *spin-echo* sequences: the circulating blood flows out of the slice; therefore it is not exposed to the selective 180° pulse and consequently is not refocused: it does not yield a signal. This is the phenomenon of flow void, which is at the root of "black blood" imaging (see section 1.2.1.1).

1.4.1.1.2. Choice of parameters

The strength of the vascular signal depends on:

– the velocity and type of flow;

– the length and orientation of the vessel being explored (the vascular signal will be stronger if the slice is perpendicular to the axis of the vessel);

– the parameters used for the sequence:

- the repetition time TR is often chosen to be as short as possible to enhance the contrast due to paradoxical enhancement,

- the thickness of the slice $\Delta z$ (see section 1.2.1.2.2),

- but also the *flip angle* (the saturation of the stationary tissues is faster if the flip angle is wide).

1.4.1.1.3. A 3D representation of flow: the MIP algorithm

Maximum Intensity Projection (MIP) consists of projecting into a plane the whole of the 3D volume acquired (by a 2D or 3D sequence).

The signal is analyzed throughout the volume, and the maximum intensity pixels (above a certain predefined threshold) are projected onto a 2D image.

In order to obtain a complete volumetric reconstruction, we need to operate on slices whose projection axis is incremented (by 10° to 15°).

All of the 2D images obtained by projection can be viewed sequentially in an animated loop, giving the illusion of a true 3D image which can be rotated and viewed from any angles of incidence.

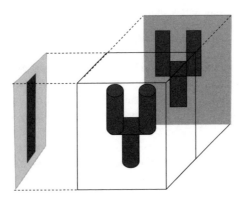

**Figure 1.24.** *Principle of MIP reconstruction*

*1.4.1.2. Limits and optimizations*

The main limitations of TOF MRA are:

– *loss of signal* when the flows are too *slow* or are oriented *in parallel* to the slice;

– *loss of signal* when the flows exhibit a *velocity gradient*: we then observe a reduction of the size of the vessels (see section 1.3.3.1); in order to deal with this problem we can use flow compensation gradients;

– *poor elimination of the signal from the stationary tissues with a short $T_1$* (primarily fat but also hematomas and thrombi), which maintain a relatively strong signal even with short TR times: they may, in certain cases, have a signal similar to that of a circulation flow lesion such as an aneurism, which skews the diagnosis.

*With regard to fat, we can then*:

– use sequences that are able to selectively eliminate the signal from the fat (such as *STIR* or *FLAIR* sequences);

– use a *magnetization-transfer* preparatory pulse: this technique is based on the fact that free protons and bound protons (belonging to macro-molecules such as fats or proteins) do not have the same resonance frequency (there is a difference of 1500 Hz). Thus, if we use a preliminary sequence that saturates the bound protons (choice of the associated frequency for the excitation pulse), those protons transfer their magnetization and thereby saturate the free protons in turn. White and gray substances, with a high concentration of bound protons, will therefore only emit a relatively weak signal. Conversely, the blood, with a low concentration of bound protons, will emit a strong signal.

*With regard to hematomas or thrombi,* there is no real "remedy": the best we can do is to favor the use of an MRA sequence in phase contrast (see section 1.4.2):

– *artifacts in MIP and placement of the acquisition volume.*

The MIP algorithm has a *threshold* below which the signal is considered null and is therefore not projected. Thus, vessels of small diameter, whose signal is weaker than the required threshold, are not visualized by projection. In this case, analysis of the slices acquired initially (the "native slices") is necessary so as not to miss any vessels.

Another artifact is due to the *placement of the acquisition volume*: when the volume is badly positioned in relation to the vessels needing to be explored, one or more vessels may be situated outside of the volume acquired. This gives us an image of false stenosis or false thrombosis on MIPs.

In order to detect this type of artifact, it is necessary to visualize projections obtained at different angles (e.g. a coronal and a sagittal approach).

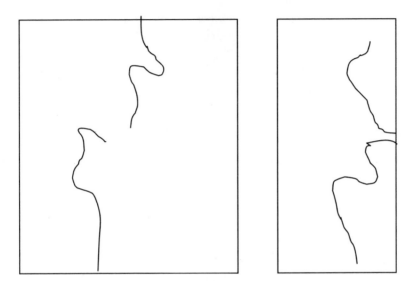

**Figure 1.25.** *Defective MIP reconstruction detected by visualization of a coronal projection (left) and a sagittal projection (right): the acquisition box is located too far to the rear, and generates a false thrombosis image*

### 1.4.1.3. Different types of acquisition of a volume

When we wish to use MRA, there is an almost automatic necessity to acquire a volume (to be aware of the vascularization of an area) and therefore not to limit ourselves to a single slice.

There are two possibilities for acquiring a volume: 2D or 3D acquisition.

### 1.4.1.3.1. 2D TOF (Time Of Flight) sequence

In 2D acquisition, TOF imaging is performed using a series of slices acquired one after another, and stacked together to reconstruct a 3D volume.

For better contrast, the slices need to be placed perpendicularly to the main trajectory of the vessel(s).

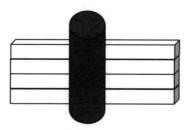

**Figure 1.26.** *Optimal relative positions of slices acquired in 2D TOF imaging and of the vessel*

*Advantages*

– Relatively fast acquisition (5 to 10 minutes, depending on the resolution and the thickness of the volume being imaged).

– Apt for different types of flows, even with slow flows (even in this case, it is possible to achieve a relatively good contrast).

– Possibility of using wide flip angles (which gives better saturation of the stationary tissues and a stronger vascular signal).

**Figure 1.27.** *Signal loss in a portion of the peroneal artery, suggesting a blockage. (From [VOS 98]. Reproduced with kind permission from Elsevier)*

*Disadvantages*

– Poor spatial resolution on the axis of stacking of the slices, because the slices are fairly thick (d > 2mm);

– The poor resolution of the voxels on the axis of stacking causes a loss of signal in areas where the flow is turbulent (wide dispersion of phases).

The same problems arise when the flows are oriented in parallel to the slice: the size of the vessels is reduced, and stenoses (which are visible by the absence of signal) may therefore be overestimated.

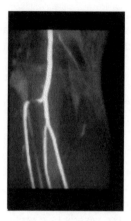

**Figure 1.28.** *Loss of signal in a horizontal portion of the right anterior tibial artery because of saturation of the flow parallel to the slice. (From [VOS 98]. Reproduced with kind permission from Elsevier)*

NOTE.– this problem is worsened in MIP imaging. Indeed, a weak vascular signal (below the specific threshold of the projection) will not appear, which causes a danger of an even greater overestimation of any stenosis.

1.4.1.3.2. 3D TOF sequence

In 3D acquisition, the whole of the stack of slices is acquired at the same time.

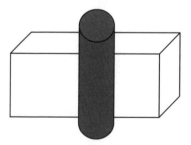

**Figure 1.29.** *Optimal relative positions of the slices acquired in 3D TOF imaging and of the vessel*

*Advantages*

– Better spatial resolution than in 2D mode: good visualization of small vessels.

– Good SNR.

– Good visualization of vessels containing areas of physiological turbulence (such as bifurcations) in comparison to 2D acquisition.

NOTE.– the problem of overestimation of stenoses, although it is less than with 2D TOF.

*Disadvantages*

– Longer acquisition times.

– Because of the excitation of a volume at each repetition, there is a *progressive saturation* of the vascular protons on the last slices downstream of the vessels, all the more so if they are slow-moving and the volume selected is thick. Thus, there is a reduction in the diameter of the vessels (see section 1.2.1.2.3). The slowest flows may even disappear completely. This type of acquisition is therefore not appropriate for the imaging of very slow flows.

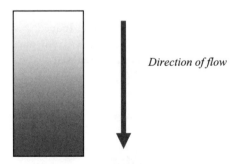

Direction of flow

**Figure 1.30.** *Attenuation of the signal from the flow because of saturation in 3D TOF. A high level of gray indicates a weak signal*

*Solutions*

In order to deal with this problem, we can decrease the saturation of the flows as they flow through the target volume by:

– Dividing the 3D acquisition into multiple blocks (or slabs).

These sequences are referred to as *MOTSA* (Multiple Overlapping Thin Slab Acquisition) sequences.

This technique has the advantage of combining good resolution (comparable to 3D imaging) whilst considerably reducing the saturation of the moving protons.

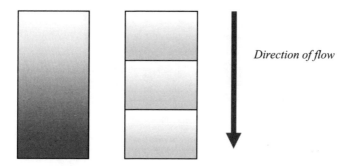

*Direction of flow*

**Figure 1.31.** *Comparison of attenuation of flow signals between in a "conventional" 3D sequence (left) and in a MOTSA sequence (right)*

NOTE.– It is necessary to cross-check adjacent volumes (overlapping) in order to prevent artifacts due to imperfections in the slice profile (the angle of excitation at the edge of the slice is less than at the center). If this overlapping is not done, we see the emergence of an artifact known as a "Venetian blind" effect.

**Figure 1.32.** *Venetian blind artifact*

– Using a variable excitation angle, smaller at the entry of the flow into the volume and greater near the exit from the volume (*TONE*: Tilted Optimized Nonsaturating Excitation), we can reduce the progressive saturation of the vessels (this also enables us to decrease the signal from the tissues with a short $T_1$, thereby obtaining better contrast).

The choice of increment of the flip angle depends on the direction and velocity of the flow and on the thickness of the acquisition volume.

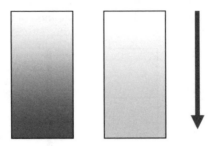

**Figure 1.33.** *Compared attenuation of flow signals in a
"conventional" 3D sequence (left) and in a TONE sequence (right)*

## 1.4.1.3.3.  2D / 3D "match"

| 2D TOF | 3D TOF |
|---|---|
| Optimal paradoxical enhancement *Choice of orientation of slices, $T_R$ and flip angle* Good contrast (because of good saturation of stationary tissues) Sensitive to slow and fast flow | Non-optimal paradoxical enhancement: progressive saturation of the vessels. *Solutions: TONE, MOTSA* Poorer contrast Sensitive mainly to fast flow |
| Poor spatial resolution: mediocre MIP reconstruction Good visualization of large vessels with unidirectional flow | High spatial resolution: good MIP reconstruction Good visualization of small vessels |
| Low SNR | High SNR |
| Short acquisition time | Long acquisition time |
| Very sensitive to turbulent flows (significant overestimation of stenoses) *Solution: Flow compensation gradients* | Less sensitive to complex and turbulent flows (slight overestimation of stenoses) |
| Poor elimination of tissues with a short $T_1$ (fat, thrombi, hematomas, etc.) *Solution for hematomas or thrombi: phase-contrast MRA (PC-MRA)* ||

## 1.4.2. *Phase-contrast angiography (PCA)*

### 1.4.2.1. *Technique*

#### 1.4.2.1.1. Principle

Phase-contrast angiography is based on dephasing of moving spins subjected to a bipolar gradient.

Remember that in a bipolar gradient, the dephasing that the spins will experience when they move along the axis of the gradient is given by: $\varphi = -\gamma \cdot v \cdot A \cdot \tau$

where $v$ is the velocity of the spins *in the direction of the gradient*, $A$ and $\tau$ are respectively the amplitude and time separating the two lobes of the gradient.

The sequences used are *gradient-echo*-type sequences.

We use a double acquisition:

– An acquisition with a *first* bipolar gradient $(+A/-A)$

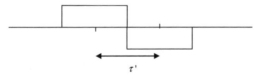

– For stationary protons, the phase obtained is equal to $\varphi^1_{statio} = \varphi_B$, with $\varphi_B$ representing the phase due to inhomogeneities.

– For moving protons, the phase obtained is: $\varphi^1_{moving} = -\gamma \cdot v \cdot A \cdot \tau + \varphi_B$

– An acquisition with a *second* bipolar gradient, of inverse polarity to the first ( $+A/-A$ ).

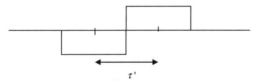

– For stationary protons, the phase obtained is equal to: $\varphi^2_{statio} = \varphi_B$

The phase due to the inhomogeneities in the field is the same as for the first gradient (same environment for the stationary protons during the application of both gradients).

 – For moving protons, the phase obtained is: $\phi^2_{moving} = +\gamma \cdot v \cdot A \cdot \tau + \phi_B$

Then, on a pixel-by-pixel basis, we subtract the signal obtained with the second gradient from the signal obtained with the first.

 – For stationary protons, the signal obtained is thus equal to:

$$S_{statio} = M \cdot e^{i\phi^1_{statio}} - M \cdot e^{i\phi^2_{statio}} = 0$$

 – For moving protons, the phase obtained is:

$$S_{moving} = M \cdot e^{i\phi^1_{moving}} - M \cdot e^{i\phi^2_{moving}} = 2M \cdot e^{i\phi_B} \cdot \sin(\gamma \cdot v \cdot A \cdot \tau)$$

*Conclusions*

 – *This subtraction eliminates the background from the image* (i.e. the stationary tissues for which the signal is null) and gives us an image where *the signal is sensitive to the velocity* $V$ of spins moving *in the direction of the gradient*: this treatment is known as velocity coding.

 – This coding is made possible even in the presence of inhomogeneities of the magnetic field ($\varphi_B \neq 0$); this demonstrates the advantage to performing two acquisitions instead of one.

 – In cases where a phase term due to the Foucault current appears, this method is no longer valid. Indeed, the phases acquired due to these currents in the first and second gradients ($\varphi_e$ and $\varphi'_e$) are different: with the pixel-by-pixel subtraction, the signal from the stationary tissues will no longer be eliminated.

 – The only option, in this case, to prevent disruption by these currents, is an appropriate antenna setup.

1.4.2.1.2. Method

In practical terms, if we wish to study the different flows in all directions in space, the method set out above requires *six* measurements (two measurements in each direction of flow coding), which is a relatively *long* process.

At present (thanks to the knowledge obtained on velocity-insensitive forms of gradient – see section 1.2.2.3), in practice the technique is modified, so that only *four* sequences are acquired:

 – we perform *one* acquisition with flow encoding gradients in *each* of the three spatial directions (i.e. 3 acquisitions): we obtain the signals $s_x$, $s_y$ and $s_z$;

– we perform an additional acquisition with a flow compensation gradient (i.e. a flow-insensitive acquisition) to obtain the reference signal $S_{ref}$.

Once the acquisitions have been made, we may then wish either to obtain the vascular anatomy, or perform measurements of velocity (and therefore flowrate), which do not involve the same data-processing procedures, as we shall now see.

*Imaging of vascular anatomy (angio-MRI)*

In the same manner as in i) above, we perform subtraction of the signals for each of the directions: $S_x - S_{ref}$ ; $S_y - S_{ref}$ ; $S_z - S_{ref}$.

These various complex subtractions are null for stationary protons and non-null for protons moving *in the directions under examination*.

We add together these three complex signals, and from the result we take the modulus:

$$S_{anat} = \left| \left( S_x - S_{ref} \right) + \left( S_y - S_{ref} \right) + \left( S_z - S_{ref} \right) \right|$$

$S_{anat}$ is non-null if there is a flow, *regardless of its direction*.

To create a 3D venous map, $S_{anat}$ is analyzed with an MIP algorithm.

NOTE.– It is of course possible to take, for instance, the modulus $\left| S_x - S_{ref} \right|$ without worrying about the other directions.

*Measuring the velocity (quantifying the flow)*

This time, it is the *phases* of the signals rather than the signals themselves that are subtracted, again for each of the directions:

$$\varphi(S_x) - \varphi(S_{ref}) \; ; \; \varphi(S_y) - \varphi(S_{ref}) \; ; \; \varphi(S_z) - \varphi(S_{ref})$$

We can then obtain a *vector* proportional to the velocity vector:

$$\overline{S_{velocity}} = \left[ \phi(S_x) - \phi(S_{ref}) \right] \cdot \vec{i} + \left[ \phi(S_y) - \phi(S_{ref}) \right] \cdot \vec{j} + \left[ \phi(S_z) - \phi(S_{ref}) \right] \cdot \vec{k}$$

$$\overline{S_{velocity}} = -\gamma \cdot A_x \cdot \tau_x \cdot v_x \cdot \vec{i} - \gamma \cdot A_y \cdot \tau_y \cdot v_y \cdot \vec{j} - \gamma \cdot A_z \cdot \tau_z \cdot v_z \cdot \vec{k}$$

NOTE.– As we saw in section 1.4.2.1.1.), the phase due to any inhomogeneities in the field is eliminated by subtraction.

We can thus establish a map with grayscale coding representing the norm of velocity at each point, and its orientation: flows running towards the tester are coded in shades of black (phase between -180° and 0°), and those running away from the probe are coded in shades of white (phase between 0° and 180°).

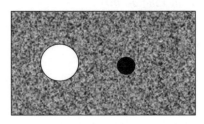

**Figure 1.34.** *Images of two flows of opposing directions in phase-contrast imaging*

Generally, outside of blood vessels, the image has a series of black and white dots representing background noise: when the two initial phases are subtracted (e.g. $\phi(S_x)$ and $\phi(S_{ref})$), as the noise factors associated with each of them are not correlated, the signal $S_{velocity}$ has low intensity and a random sign outside of the vessels. This background noise can be eliminated if a threshold level for the value of the phase is imposed on the final image: the background then appears gray (i.e. an intermediary color).

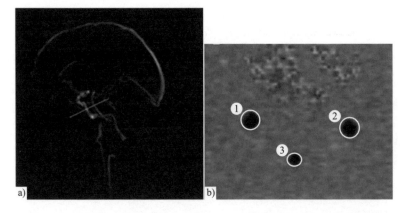

**Figure 1.35.** *Phase-contrast imaging. Top: sagittal slice of location. Bottom: 2D PCA done in the slice marked by a line on the location image, perpendicular to the carotid and basilar arteries. 1 and 2: carotid arteries, and 3: basilar artery*

*1.4.2.2. Choice of parameters and artifacts relating to the sequence*

One of the main parameters needing to be determined in phase-contrast MRA is the *velocity encoding* $V_{enc}$. This is determined by the area $A \cdot \tau$ of the bipolar gradient, and is defined by the velocity of the protons giving a dephasing of 180° ($\pi$ radians):

$$\pi = -\gamma \cdot V_{enc} \cdot A \cdot \tau$$

The choice of velocity encoding is of crucial importance; if the velocity encoding is not at least equal to or greater than the maximum velocity of the protons in the vessels that we wish to image, a phenomenon of *aliasing of the velocities* occurs.

Hence, let us consider an velocity encoding equal to 50 cm/s; the protons circulating at this velocity have a phase of 180°.

Fast-moving protons, e.g. moving at 75 cm/s, will exhibit 270° dephasing.

Slow-moving protons, e.g. moving at 25 cm/s, and in the *opposite orientation* to the previous set of protons, will have -90° dephasing.

These two types of protons cannot be differentiated from one another, and are therefore coding with the same level of gray: this is the phenomenon of aliasing.

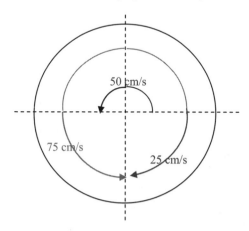

**Figure 1.36.** *Phenomenon of aliasing in PCA flow imaging*

The characteristics of the encoding gradient are therefore defined in order to be able to encode flows within a certain range of velocity: between $-V_{enc}$ and $+V_{enc}$. Any velocity beyond this range will be incorrectly encoded.

In order to avoid aliasing, we need to estimate the highest velocities in the vessels which we wish to image.

In practice, velocity aliasing is not massively problematic, because the flow velocities are usually greater in the center of the vessel (this is known as a laminar flow): the vessel will be clearly delimited by the protons circulating at a lower velocity at the edges.

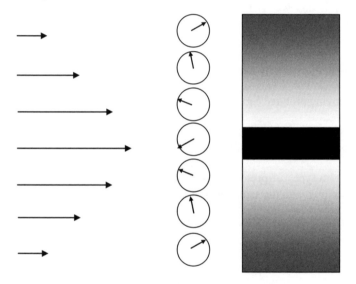

**Figure 1.37.** *Phenomenon of aliasing observed at the center of the vessel (the velocity is too great in comparison to the velocity encoding, so we have a phase between -180° and 0°). The contours of the vessel, though, are correctly imaged*

Before launching a sequence, we need to test various velocity encodings in order to choose the one which is most appropriate (this is what is generally done with a 2D PCA sequence: see below); it should be noted that the velocity encoding may be different in each spatial direction.

*Example*

A velocity encoding of 20 to 40 cm/s is indeed appropriate for venous flow.

A velocity encoding of 60 to 80 cm/s is appropriate to visualize the cerebral arteries (to the detriment of slow flows such as venous flow, whose signal would be very weak).

*Other parameters relating to the method*

Similarly to TOF-MRA, the $T_R$ is chosen to be short (30-100ms), but care is taken not to cause the saturation of the vessels in the slice of interest.

The limitations of this technique relate to the loss of signal caused by complex or turbulent flows that is responsible for intravoxel dephasings: in order to reduce this artifact, the echo time $T_E$ will be chosen as short as possible (8-14ms).

This loss of signal is all the greater when the voxel is large; in 3D imaging, therefore, we always try to choose a small voxel size.

Finally, the *artifact of MIP and placement of the acquisition volume*, which we saw above in our discussion of TOF sequences, is also present with the PCA sequence.

### 1.4.2.3. *Different types of acquisition of a volume*

#### 1.4.2.3.1. 2D PCA sequence

The advantages to 2D PCA are several:

– its rapidity (a slice can be acquired in under two minutes), which means we can test different encoding velocities for a possible 3D acquisition;

– the possibility of creating cinematic imaging of the vascular flows (with ECG synchronization);

– the option to plot a curve showing the velocity over time within the same slice. This velocity curve is then coupled with a morphological image to give us the breadth of the vessels and deduce the vascular flowrate.

#### 1.4.2.3.2. 3D PCA sequence

The main advantages to 3D PCA over 2D PCA are similar to those of 3D TOF over 2D TOF:

– much better resolution, so less sensitive to turbulent flows;

– improved SNR.

The main disadvantage of the sequence is its relatively long acquisition time (over 10 minutes); this time is twice as great as for 3D TOF imaging (because twice as many acquisitions are made). In an attempt to reduce it, we can decrease the number of phase encoding steps, or use parallel imaging (*SENSE*).

| 2D PCA | 3D PCA |
|---|---|
| – Very good elimination of stationary tissues (even those with short $T_1$ times) – Very good contrast | |
| – Poor spatial resolution – No MIP reconstruction | – Better spatial resolution – Good quality MIP reconstruction |
| – Sensitive to slow and fast flows | |
| – Short acquisition time | – Very long acquisition times |
| – Very sensitive to complex and turbulent flows (overestimation of stenoses) | – Less sensitive to complex and turbulent flows (less than 2D PCA but more than 3D TOF) (slight overestimation of stenoses) |

### 1.4.3. *The "match" between TOF and PCA imaging*

The table below illustrates the advantages of one method over another.

| TOF | PCA |
|---|---|
| – Appropriate for vessels with slow to fast flow | – Appropriate for vessels with very slow to fast flow |
| – Better visualization of the vessels containing physiological turbulences | |
| – Shorter acquisition times with identical spatial resolution | |
| – Better spatial resolution with identical acquisition times | |
| | – Better contrast because of better elimination of stationary tissues |

*Conclusion to this table:* the choice between TOF and PCA is dictated by the region needing to be explored and the type of bloodflow: 3D TOF offers the best performance for the exploration of intracranial vessels (good resolution needed, fast flow); 2D- or 3D PCA sequences are better adapted for studying the cerebral veins (slow flows).

### 1.4.4. *Contrast-enhanced MRA (CE-MRA)*

With the two techniques discussed above (TOF and PCA), MRA has taken off and has become part of medical routine, particularly in the study of the brain.

However, there are three main criticisms that can be leveled at these techniques:

– the acquisition times are particularly long;

– the issues of the protons' saturation mean that the volumes which can be explored are small (i.e. limited FOVs, fields of view);

– the temporal resolution obtained with these methods is very poor (there is no possibility of dynamic tracking of the flow).

With a view to getting around these difficulties, a new technique has been developed: CE-MRA.

#### 1.4.4.1. *Technique*

1.4.4.1.1. Principle and advantages of the sequence

CE-MRA is an ultra-fast 3D gradient-echo sequence with a spoiler gradient placed after the acquisition to counter the residual magnetization and obtain significant weighting $T_1$.

The technique involves the injection of a paramagnetic substance: gadolinium (Gd).

The gadolinium creates an intense local magnetic field which disturbs the environment of the protons of the medium in which it is: thus, their relaxation is accelerated, and the Gd artificially lessens the $T_1$ of the blood (and the $T_2$, but only with high doses).

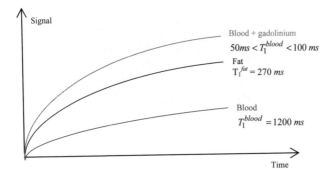

**Figure 1.38.** *Compared values for $T_1$ and relaxations of blood (with and without gadolinium) and fat in a field $B_o = 1.5\,T$*

Thus, the vascular signal obtained with this sequence is *intense*, and *relatively independent of the flow artifacts* discussed above:

– no saturation of the blood signal, which enables us to explore large volumes;

– little loss of signal relating to intravoxel dephasing, which means we can image turbulent flows.

NOTE.– There is always intravoxel dephasing, but the consequent decrease in the $T_1$ "masks" the effects of this dephasing.

In addition, CE-MRA is set apart from PCA and TOF sequences by virtue of the fact that:

– it is faster: the acquisition is done in the space of a few seconds;

– it provides better contrast: the stationary protons are saturated thanks to the short $T_R$ times;

– it decreases the motion-related artifacts: there is the option of performing sequences under controlled apnea (so there is less sensitivity to these artifacts).

This form of acquisition, after treatment (MIP algorithm, as used with TOF or PCA sequences), yields images similar to those obtained by "conventional" angiography, with the advantage of not being irradiant.

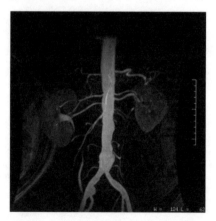

**Figure 1.39.** *CE-MRA of the abdominal aorta and its renal branches. (From [AUE 04]. Reproduced with kind permission from Elsevier)*

1.4.4.1.2. Description of the k-space and injection of gadolinium

We know that the central rows in the k-space play a part in the contrast of the image and the SNR, whereas the peripheral rows determine the spatial resolution.

Thus, in order to obtain as high a contrast as possible, the center of the k-space needs to be recorded when the concentration of gadolinium (when it first passes through) in the slice is maximal (maximal signal).

This concentration peak is very fleeting, so we need to:

– *choose a good definition of the k-space* which favors a quick start to acquisition, enabling the center of the space to be acquired rapidly:

- row-by-row definition (in a linear or sequential fashion). Creation of the contrast of the image requires 20% of the acquisition time,

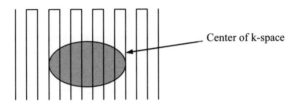

**Figure 1.40.** *Cartesian definition of k-space*

- elliptical-central definition: the center of the k-space is obtained first by creating an ellipse, which is equivalent to acquiring the data by a range of frequencies (from the lowest to the highest). Creation of the contrast of the image now requires no more than 2% of the acquisition time. Yet the image quality may be worse (overall blur) than with row-by-row scanning if the gradients are not properly stabilized;

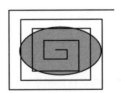

**Figure 1.41.** *Elliptical-central definition of k-space*

– *know the transit time of the gadolinium* from the injection point to the vessels being imaged. This time corresponds to an acquisition window between 5 and 20 seconds (for the slowest flows).

A variety of methods are available to perform good synchronization:

– carry out a test injection and acquire a slice centered on the region of interest every second for a minute. A region of interest (ROI) enables us to determine the transit time by pinpointing the peak of the signal. This method is relatively slow but it is the most reliable;

– perform automatic detection: a volume of interest (VOI or tracker) is positioned upstream of the vessel needing to be explored;

– a TSE (Turbo Spin Echo) sequence automatically analyzes the intensity of the signal in the vessel. When the threshold is reached, the CE-MRA sequence is triggered, either automatically or after a fixed delay;

– the main drawback to this method is the danger of incorrect analysis of the intensity of the signal, polluted by motion-related artifacts, for instance;

– semi-automatic detection: real-time visualization of the arrival of the gadolinium in the vessels with an ultra-fast (1FPS) 2D gradient-echo dynamic sequence.

### 1.4.4.2. *Parameters relating to the sequence*

All the usual criteria used in a "conventional" sequence need to be reviewed, bearing in mind that we want to make a 3D acquisition very fast with a wide FOV.

#### 1.4.4.2.1. Repetition time, echo time and sweep width

These two times are chosen to be *as short as possible* so as to decrease the total acquisition time: $TR \leq 5\ ms$ and $TE \leq 2\ ms$ .

It is not helpful to compensate the flow (see section 1.2.2.3): indeed, the gain in $T_1$ masks all the flow-related phenomena. Furthermore, any compensation would prolong the $TE$ .

NOTE.– With a <u>CE-MRA</u> sequence, it is possible to decrease the echo time. Indeed:

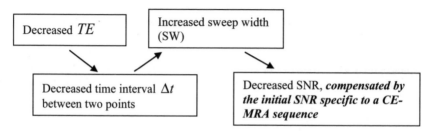

It is also possible to use specific echo times (2.2 ms at 1.5 T) where the fat and water are either in phase or in phase opposition, which enables us to remove the image of the fat.

## 1.4.4.2.2. FOV

The choice of the orientation of the volume to be imaged is entirely independent of the orientation of the vessels (unlike TOF and PCA) because the phenomenon of saturation is no longer present.

With CE-MRA, it is possible to obtain extended FOVs: thus, we can image the abdomen, the lower limbs, etc.

*A compromise*

If the FOV is widened and the size of the matrix remains the same, the spatial resolution suffers; if we wish to preserve the same spatial resolution, it is necessary to increase the acquisition time.

A compromise needs to be found between acquisition time, FOV, spatial resolution and SNR.

*Solutions*

– Generally, the FOV and the matrix are rectangular (as they are in "classic" imaging).

– We can perform zero-filling in the direction of the slice selection gradient (which reduces the acquisition time if the patient moves too much, for instance).

– Even with the two solutions suggested above, the acquisition time remains relatively long and the spatial resolution is sub-optimal; this is true even with the best technical characteristics of the gradients (which determine the acquisition times).

Parallel acquisition techniques, which deliberately cause aliasing of the image (in the k-space with *SMASH* and in the real domain with *SENSE*) help to considerably reduce the acquisition time or significantly increase the spatial resolution without placing further constraints on the gradients.

In addition, CE-MRA has an intrinsic SNR which is sufficient to compensate for the loss due to the use of these techniques.

*Supra-aortic angio-MRI uses these techniques. Venous angio-MRI does not, as time-gain is not crucially important in this application.*

Generally, the spatial resolution in CE-MRA is poorer than that achieved with 3D TOF for studying intracranial vessels.

1.4.4.2.3. Calculating the injection delay

The passage of the gadolinium through the slice needs to coincide with the acquisition of the central rows in the Fourier plot. A precise timing therefore needs to be established, giving the delay $T_{dem}$ between the injection and the start of acquisition.

For this purpose, it is useful to describe the evolution of the signal over time that is characteristic of its evolution.

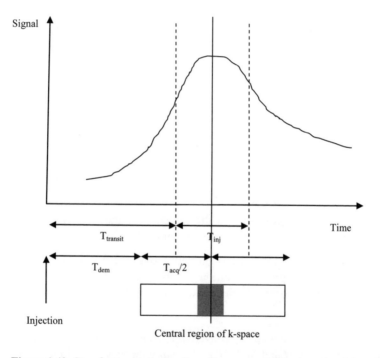

**Figure 1.42.** *Signal intensity in the slice of interest and injection of gadolinium*

*Criteria and parameters needing to be respected*

The duration of infusion of the bolus $T_{inj}$ (i.e. the time period over which the gadolinium injection is administered) needs to cover at least half the acquisition time $T_{acq}$ so that the central rows of the k-space can be acquired with maximum signal strength:

– the transit time $T_{transit}$ can vary between 6 and 40 seconds: this time can be evaluated with an injection test;

– generally, in order to improve the visibility of the images obtained, we need to choose between arterial and venous imaging.

This choice can be made on the basis of the period when the acquisition needs to take place: if we wish to avoid venous contamination, the acquisition needs to be made during the first phase of the gadolinium's arterial passage.

In the case of cerebral imaging, for instance, the acquisition time $T_{acq}$ needs to be very short (10-12 seconds) which is the approximate length of time taken for blood to complete the cerebral arterial circuit.

Given these different times, referring to the curve giving the intensity of the signal in the slice over time, we have:

$$T_{dem} = T_{transit} + T_{inj}/2 - T_{acq}/2$$

1.4.4.2.4. Flip angle

There is less of an impact on the vascular contrast than with PCA and especially TOF. Habitually, the flip angle used for arterial imaging is between 50° and 60°, and for venous imaging between 30° and 40°.

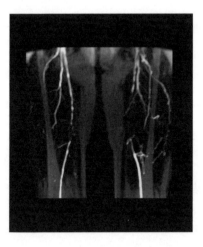

**Figure 1.43.** *CE-MRA showing a bilateral occlusion of the superficial femoral artery. (From [AUE 04]. Reproduced with kind permission from Elsevier)*

1.4.4.3. *Contrast-enhancing strategies*

1.4.4.3.1. Before acquisition of the sequence

Preparatory sequences (STIR or FLAIR) can be integrated with the CE-MRA sequence. These help to reduce the signal from the stationary tissues, although they do not eliminate it entirely.

On the other hand, they do lead to a consequent increase in acquisition time.

1.4.4.3.2. After acquisition of the sequence

The taking of an acquisition before the bolus arrives (see Figure 1.38) can be used to perform a *subtraction*, as done with the PCA sequence, to eliminate the stationary tissues, including those with short $T_1$ times.

This subtraction is crucial in order to view small structures such as the intracranial vessels.

However, as it requires the patient to remain perfectly still, subtraction is not used in regions where the movements (heart pulses, breathing movements, etc.) are significant, such as in the neck.

1.4.4.4. *Artifacts*

1.4.4.4.1. Artifact of timing

As we saw above, the speed of the CE-MRA sequence and the fleeting passage of the gadolinium through the slice of interest necessitate very precise timing.

If this condition is not respected to the letter, certain artifacts may arise.

In particular, when the acquisition is triggered too late, a venous signal appears which "pollutes" the image, rendering it less clear.

1.4.4.4.2. Artifact of MIP and placement of the acquisition volume

This artifact has already been discussed with regard to the TOF and PCA methods, but is also observable with CE-MRA.

*1.4.4.5. The positives (+) and negatives (–) of the CE-MRA technique*

| + |
|---|
| • Wide FOV |
| • Excellent elimination of stationary tissues (subtraction technique) |
| • Short acquisition time (20-40s) |
| • Less sensitivity to movements |
| • Fairly good temporal resolution (but greatly inferior to that achieved with "conventional" angiography, which is 0.5 s) |
| Independent of flow-related phenomena <br> • Good spatial resolution (little intravoxel dephasing) <br> • Not hugely sensitive to turbulent flows |
| – |
| • More invasive <br> • Higher cost <br> • Negative effect on the $T_2$* of the gadolinium at high doses |

NOTE.– With the aim of further improving the temporal resolution of CE-MRA, it is possible to perform dynamic angio-MRI: the 3D sequences obtained are optimized and the temporal resolution is around 1.4 s, which is still less than that obtained with conventional angiography.

## 1.5. Study of two clinical problems

The choice of which type of sequence to use to image the flow, from amongst the various options advanced in this chapter, must be based on the type of disease from which it is suspected that the patient is suffering.

Indeed, if we choose an "incorrect" sequence, there is a danger of causing confusion and artificially aggravating the problems detected.

### 1.5.1. *Differentiation between a thrombus and a slow flow*

#### 1.5.1.1. *The problem*

The distinction between a thrombus (a mass of coagulated blood that is formed in a vessel) and a slow flow is a problem that is regularly encountered in clinical routine.

In the case of a conventional TOF sequence, the signals from a thrombus and a slow flow are seemingly the same, because:

– the slow flow experiences a slight phenomenon of paradoxical enhancement;

– the thrombus has a fairly short $T_1$ and a relatively high $T_2^*$ (owing to the presence of methemoglobin, which has a paramagnetic effect).

How, in this case, can we differentiate a thrombus from a slow flow?

### 1.5.1.2. *Proposals and solutions*

A number of strategies are put forward below. For each of them, we indicate whether it could indeed be a solution to the problem at hand, and the reasons for this judgment.

#### – Increasing the TE?                             → YES

This increase in TE leads to a significant weighting $T_2$ (or $T_2^*$) of the MRA sequences:

- the thrombus has a relatively strong signal regardless of the value of the TE;

- the signal from the slow flow will fade.

#### – Observation of artifacts?              → YES

If a 3D acquisition is performed, is there a *decrease in the signal in the direction of the flow* within the volume? If so, we are witnessing a slow flow.

Is there an *artifact of ghosting* near to the vessel? If so, we are looking at a slow flow.

#### – Injection of gadolinium?              → NO

This injection cannot solve the problem, because the signals from a slow flow and from a thrombus will both be increased.

#### – Application of a pre-saturation slab?    → YES

This strip needs to be placed above or below the structure that is suspected of being a thrombus, in order to validate or debunk this hypothesis: if the signal is reduced, then we are dealing with a slow flow.

#### – Use of PCA?                              → YES and NO

This is possible if the velocity encoding is sufficiently low to enable us to see slow flows. The signal from the thrombus still should not appear. However, take care: extremely slow flows will still not appear.

**– Two acquisitions with TOF sequences?    → YES and NO**

As we saw above, a single TOF sequence is not enough. Two TOF sequences might alleviate the doubt if:

- those two sequences are acquired with slices oriented in different directions;

- one of the two sequences has a flow compensation gradient and the other does not.

In both cases, the signal from a slow flow will be altered, but the signal from a thrombus will not.

### 1.5.2. *"Correct" evaluation of a stenosis*

#### 1.5.2.1. *Problem*

The correct evaluation of the diameter of a vessel is made difficult by the loss of signal at the periphery of that vessel engendered by the phase heterogeneity of the protons within the same voxel (see section 1.3.3.1).

In order to deal with this problem, we saw in our discussion of flows at constant velocity that it was possible to use a sequence with flow compensation gradients.

The case of stenosis (constriction of the caliber of a vessel) is different: it is not possible to rephase the signal with a flow compensation gradient because the acceleration of the blood is non-null.

#### 1.5.2.2. *Possible solution*

The above problem stems from the fact that the sequences used (with a flow compensation gradient) show flow with constant velocity as a hypersignal and stationary tissues as a hyposignal.

Hence, exhibiting acceleration is confused with the stationary tissues.

One possible solution is to image the stationary tissues as a hypersignal and "the rest" (flow with any velocity, be it constant or otherwise) as a hyposignal.

The sequence used for this type of imaging is a *black-blood TOF* sequence.

# Chapter 2

# Diffusion

This chapter is constructed in the form of a problem posed, which we are seeking to resolve (section 2.1): diffusion may help us to detect certain diseases; what can we do to make it "visible" on an MRI image?

We begin by giving one suggestion for a sequence (section 2.2).

The signal obtained with this type of sequence is then calculated mathematically (section 2.3), which enables us to demonstrate, in addition to the influence of the diffusion (diffusion coefficient, section 2.3.2), the influence of other factors such as the gradients (b factor), or relaxation time T2.

The diffusion sequence is then finalized (section 2.4) and the associated images presented, with a critique of their result in light of the mathematical signal obtained in section 2.2.

## 2.1. General points

### 2.1.1. *What is diffusion?*

In a biological medium, the molecules of water are not immobile: they are subject to permanent agitation which is known as "Brownian motion". These molecular motions are random and will be more or less intense depending on the medium: they characterize molecular diffusion.

If we pour ink onto a sheet of newspaper, the center of the mark will move if the flow is non-null. This movement can be described by a vector, which gives the direction, the orientation and the velocity of the flow.

The flow indicates a coherent overall motion.

If we now pour ink into a glass filled with water, the ink diffuses throughout the medium: the shape of the ink as it gradually becomes distributed throughout the glass is a sphere. If the flow is null, the center of the mark will not move.

| | Molecular motion | Vectors describing the motion |
|---|---|---|
| Flow | | |
| Isotropic diffusion | | |

Hence, the phenomena of flow and diffusion are entirely independent.

### 2.1.2. What is the medical interest held by diffusion?

As we shall see later on, it is the movements of the extracellular water that are explored by diffusion imaging. This provides indirect information about the structure surrounding these water molecules. In the body, the movements of diffusion encounter different obstacles, which vary depending on the tissues involved, and certain disease-induced alterations (edema, abscesses, tumors, etc.). Hence, diffusion can aid in the detection of these diseases.

### 2.1.3. The three main types of diffusion

We can identify three main types of diffusive motions: free, isotropic restricted or anisotropic restricted.

*Free diffusion*

There are no obstacles facing the movement of the water molecules. This type of motion is possible in cerebrospinal fluid (CSF), for instance.

**Figure 2.1.** *Trajectory of water molecules in the case of free diffusion*

*Isotropic restricted diffusion*

In all spatial directions, the water molecules have limited displacement. The restrictions may be pathological in nature (e.g. abscesses, tumors, etc.) or otherwise.

**Figure 2.2.** *Trajectory of water molecules in the case of isotropic restricted diffusion*

*Anisotropic restricted diffusion*

Certain tissues constitute obstacles which direct the molecules in particular directions: diffusion is only restricted in certain spatial directions.

Such is the case with nerve fibers comprising beams of axons and concentric layers of myelin (the fatty "sheath") which prevents any transversal diffusion.

**Figure 2.3.** *Trajectory of water molecules in the case of anisotropic restricted diffusion*

## 2.2. Principle behind diffusion imaging and the associated sequence

*The aim in this section is to give "hands-on" explanations; the calculations used to refine the sequence and to gain a clearer understanding of the results obtained will be seen later on.*

The objective of diffusion-weighted sequences is to obtain images whose contrast is influenced by the differences in mobility of the water molecules.

However, the "conventional" sequences are not sensitive to diffusion: in these sequences, the signal loss caused by protons moving in a voxel, in comparison to significant molecular agitation, is imperceptible (we disregard the macroscopic flow).

Thus, we add so-called "diffusion gradients" in the preparatory phase for an imaging sequence: these diffusion gradients, which are far greater (see section 2.3.3.3) than the phase-encoding- or frequency gradients, are applied to both sides of a 180° refocusing pulse.

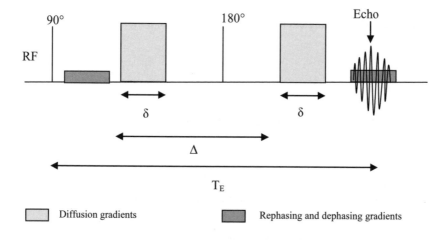

**Figure 2.4.** *Diffusion sequence*

*The so-called "dephasing" and "rephasing" gradients are the normal phase-encoding and frequency gradients.*

In this sequence, δ is the duration of the application of the diffusion gradients and Δ is the length of time between the starts of the application of these gradients.

Therefore, we need to distinguish between different categories of spins in order to know what becomes of them with this type of gradient:

– The spins of water molecules which do not move between the application of the two gradients are dephased by the first gradient and rephased by the second.

– The spins of those water molecules which are moved in the direction of the gradients during the delay between the two applications of gradients will not be rephased by the second gradient. Two scenarios need to be taken into account:

- For spins animated with a coherent motion (which is the case with flows), the dephasing caused by the diffusion gradients does not cause an overall decrease in the amplitude of the magnetization gradient.

- With random motions, such as diffusion (or perfusion: see Chapter 3), the phase dispersion leads to a decrease in the magnetization gradient and therefore an attenuation of the measured signal. The decreased amplitude of the signal depends on the proportion of spins which have diffused between the two gradients, i.e. on the diffusion coefficient.

The crucial difference highlighted here enables us to prove that the sequences which we shall use in this chapter cannot be used to image bloodflow (see Chapter 1) but rather the water-based environment of biological tissues.

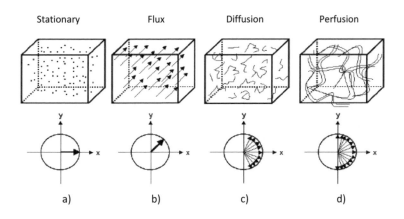

**Figure 2.5.** *Different types of spins and dephasing induced by the diffusion sequence*

In summary, diffusion-weighted images show a *hypersignal* in regions with *reduced molecular diffusion* and a weaker signal when the area being explored contains molecules exhibiting high diffusion.

## 2.3. Study of the obtained signal

The sequence seen above still needs to be added to in order to yield the image of an entire slice. We shall therefore build upon it in the following pages. For the moment, with this "beginning" of a sequence, we are able to give the shape of the signal obtained and draw additional conclusions from this.

### 2.3.1. *Expression of the signal*

2.3.1.1. *Variation in transversal magnetization over time in the absence of diffusion*

Below, we recall the Bloch equation showing the fate of transverse magnetization within a voxel in the presence of a gradient in the rotating trihedron, considering the phenomenon of relaxation $T_2$:

$$\frac{dM_T}{dt} = -j\gamma\left(\overrightarrow{G}\cdot\overrightarrow{r}\right)M_T - \frac{M_T}{T_2} \qquad [2.1]$$

2.3.1.2. *Variation in transversal magnetization over time in the presence of diffusion only*

Let us suppose that the gradients applied are null, and for the time being, omit the transversal relaxation (this means that the term $T_2$ is considered to be very large): therefore the right-hand term in the above equation will be null.

The only possible variation $dM_T$ in the transverse magnetization within a voxel would then result from the flow of magnetization from neighboring voxels. This flow is made possible by the differences in magnetization (or more specifically in density of protons) throughout the tissue.

For simplicity's sake, we shall suppose in our discussion below that diffusion can only take place in direction x.

Consider a "tube of tissue" with surface area S, formed between two abscissa points: x and x+dx.

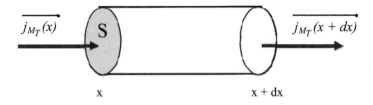

**Figure 2.6.** *Flow of magnetization through a tube of tissue*

The flow of magnetization into this slice at abscissa S will be greater when the magnetization of this slice is slight in comparison to that of the slice situated to its left ("nature abhors a vacuum"!); thus, the flow is characterized by the magnetization flow density vector $\overrightarrow{j_{M_T}}$, whose projection in direction x is

$$j_{M_T}(x) = -\frac{1}{S \cdot dx} \cdot D \cdot \frac{dM_T}{dx}\Big|_x$$

D is the diffusion coefficient and the term $\dfrac{1}{S \cdot dx}$ is included to nullify the influence of the volume of the tube.

The term "$-$" is here to ensure respect for the fact that the flow of magnetization runs from a medium where magnetization is high to one where it is low.

Similarly, we can define the magnetization flow density vector at the other end of the slice: $j_{M_T}(x+dx) = -\dfrac{1}{S \cdot dx} \cdot D \cdot \dfrac{dM_T}{dx}\Big|_{x+dx}$.

The variation $dM_T$ over a given time-period dt is therefore due to the magnetization "entering" the tube: $j_{M_T}(x) \cdot S \cdot dt$, from which we subtract the magnetization "exiting" the tube: $j_{M_T}(x+dx) \cdot S \cdot dt$.

Thus:

$$dM_T = j_{M_T}(x) \cdot S \cdot dt - j_{M_T}(x+dx) \cdot S \cdot dt$$

This gives us

$$\frac{dM_T}{dt} \cdot dx = D \cdot \left( \frac{dM_T}{dx}\Big|_x - \frac{dM_T}{dx}\Big|_{x+dx} \right)$$

Finally:

$$\frac{dM_T}{dt} = D \cdot \frac{d^2 M_T}{dx^2}\Big|_x \qquad\qquad [2.2]$$

### 2.3.1.3. *Variation in transversal magnetization over time*

We shall simultaneously now take account of the presence of a gradient, the relaxation $T_2$ and the phenomenon of diffusion: hence, we combine expressions [2.1] and [2.2]:

$$\frac{dM_T}{dt} = -j\gamma\left(\vec{G}\cdot\vec{r}\right)M_T - \frac{M_T}{T_2} + D\cdot\frac{d^2 M_T}{dx^2}\bigg|_x \qquad [2.3]$$

In order to obtain the solution to this equation, we begin with the solution to the equation without the diffusion term:

$$M_T(t) = M_T(0)\exp\left(-j\gamma x\int_0^t G_x(t)dt - \frac{t}{T_2}\right)$$

The constant term $M_T(0)$ is replaced by a time-variant term $A(t)$:

$$M_T(t) = A(t)\exp\left(-j\gamma x\int_0^t G_x(t)dt - \frac{t}{T_2}\right) \qquad [2.4]$$

We introduce this expression into equation [2.3]:

$$\frac{dA(t)}{dt} = -D\left[\gamma\int_0^t G_x(t)dt\right]^2 A(t)$$

We solve this equation and substitute the result back into equation [2.4]:

$$M_T(t) = M_T(0)\exp\left[-D\gamma^2\int_0^t\left(\int_0^\tau G_x(\tau')d\tau'\right)^2 d\tau\right]\exp\left(-j\gamma x\int_0^t G_x(t)dt - \frac{t}{T_2}\right) \qquad [2.5]$$

with $0 \leq \tau \leq t$.

We choose a layout of the k-space which favors the contrast of the image: the central rows are obtained for $t = T_E$.

This choice implies that $\int_0^{TE} G_x(\tau')d\tau' = 0$

Thus:

$$M_T(T_E) = M_T(0)\exp(-bD)\exp\left(-\frac{T_E}{T_2}\right)$$

[2.6]

with

$$b = \int\limits_0^{TE}\left(\int\limits_0^{\tau} G_x(\tau')d\tau'\right)^2 d\tau$$

*Conclusions*

As we saw in section 2.2, we note that areas with significant diffusion (a high value of D) have a weak signal.

The first part of expression [2.6] describes the attenuation of the signal due to the diffusion, and the second describes that which is due to the gradients and relaxation $T_2$.

The attenuation due to diffusion can itself be broken down into two parts: one part relating to the gradient (b) and one part relating to the tissue being imaged (D): these two components will be discussed separately below.

### 2.3.2. *Diffusion coefficient*

The diffusion coefficient D is expressed in $mm^2/s$.

In a given medium, the average diffusion distance $\Delta x$ over a time interval $\Delta t$ is: $\Delta x = \sqrt{2D\Delta t}$ : the larger the value of D, the greater the average distance travelled by the molecules.

D is a physical parameter which does not depend on the magnetic environment (unlike times $T_1$ and $T_2$).

The values of D are characteristics of the tissues at a given temperature, presented below in decreasing order:

– CSF: $D_{LCR} = 3 \cdot 10^{-3} mm^2 / s$

– gray matter:

  - cortex: $D_{cortex} = 0.89 \cdot 10^{-3} mm^2 / s$

- basal nuclei: $D_{BN} = 0.75 \cdot 10^{-3} \, mm^2 \, / \, s$

– white matter: $D_{WM} = 0.70 \cdot 10^{-3} \, mm^2 \, / \, s$

NOTE.– These coefficients are lower than that of pure water ($D_{pure\ water} = 3.2 \cdot 10^{-3} \, mm^2 \, / \, s$) because of the obstacles encountered by the water molecules in the tissues (cell membranes, molecular barriers, etc.) which slow diffusion.

### 2.3.3. The b factor

#### 2.3.3.1. Value of b and working scale

The b factor is expressed in $s/mm^2$ (the opposite unit to that used for D). It is a factor which is directly dependent upon the force of the gradients.

As we can see from the signal attenuation coefficient ($\exp(-bD)$), b can be used to determine the sensitivity of the measurement, i.e. the scale we need to use to observe the diffusion, and thus examine this diffusion in the different constituent compartments of the tissues.

In principle, though, all the random motions are observable with a diffusion sequence. What, therefore, are the different scales on which these random motions take place?

We can distinguish three such scales:

– water, within the cells (diffusion coefficient $D_{int}$);

– water, outside the cells (diffusion coefficient $D_{ext}$);

– blood, within the tissues: its tortuous trajectory and its numerous changes in direction are similar to a random motion called "perfusion" (diffusion coefficient $D_p$).

The velocity of the blood in this microcirculation is far greater than that of water in its intra- or extra-cellular movements: the mechanism of perfusion thus involves far greater diffusion coefficients: $D_p \gg D_{int}$ and $D_p \gg D_{ext}$.

However, only a fraction of each voxel is occupied by the capillary network, which decreases the apparent diffusion coefficient of the perfusion mechanism.

Ultimately, we can say that $D_p$ is around 10 times greater than $D_{int}$ or $D_{ext}$: this difference means that it is possible to separate these three scales.

Thus, if we work with a field $B_o = 1.5$ T, we estimate that for:

− b < 100 s/mm$^2$, we measure large values for the diffusion coefficients: the imaging technique associated with perfusion will later be discussed in a dedicated chapter (see Chapter 3);

− 300 < b < 3000 s/mm$^2$, the diffusion distances measured tend to correspond to extra-cellular water diffusion;

− b > 6000 s/mm$^2$, we are dealing with intra-cellular diffusion.

### 2.3.3.2. *What value should we choose for b, then?*

As we saw at the very beginning of this chapter, it is the extra-cellular water movements that we wish to study in diffusion MRI.

We have also seen that by increasing the value of b, we increase the sensitivity of the sequence to the phenomenon of diffusion.

Finally, we know that increasing the value of b leads to a decrease in the SNR.

For all of these reasons, we typically choose a value for b of around 1000 s/mm$^2$.

### 2.3.3.3. *Necessity of significant gradients*

Using the notations from section 2.2, and re-using the expression obtained previously, we can calculate the value of b and thereby obtain the well-known

Stejskal–Tanner formula: $b = \int\limits_0^{TE} \left( \int\limits_0^{\tau} G_x(\tau')d\tau' \right)^2 d\tau = \gamma^2 \delta^2 \left( \Delta - \frac{1}{3}\delta \right) G_x^2$

NOTE.– $b$ is proportional to $G_x^2$: it is necessary to use powerful gradients so that the value of b is sufficiently great, and therefore so that the sequence will be heavily diffusion-weighted.

## 2.4. Diffusion sequence and diffusion images

In view of the study carried out in section 2.3, we shall now refine the sequence given in section 2.2 and observe the images obtained with this sequence.

### 2.4.1. *Complete sequence*

In order to give the definitive form of the diffusion sequence, it is necessary to take account of the additional constraints in relation to the sequence championed in section 2.2:

– significant gradients are necessary in order to facilitate appropriate weighting of the sequence in regard to the phenomenon of diffusion;

– diffusion is even more susceptible to artifacts of motion than the other phenomena being imaged (flow, etc.); the sequence therefore needs to be as fast as possible.

The sequence chosen is an *Echo Planar Imaging – Spin Echo (EPI-SE) sequence* in which the diffusion gradients are applied in the three axes x, y and z.

The echo planar imaging sequence is indeed particularly appropriate for this type of exploration: this ultra-fast imaging technique helps reduce artifacts caused by physiological movements, and the technique uses "powerful" gradients which are necessary to achieve the values for the coefficient b seen above.

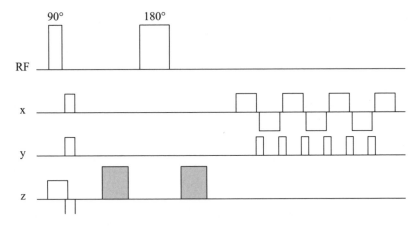

**Figure 2.7.** *Diffusion sequence with gradients on all three axes*

### 2.4.2. *Diffusion image*

#### 2.4.2.1. *Protocol*

The sensitivity of the diffusion sequence presented above is limited to diffusion only in the direction of the gradients (here in direction z).

Thus, in practice, a complete diffusion sequence requires the successive application of three EPI sequences using diffusion gradients respectively along the axis of slice selection, phase encoding and frequency encoding: in total, three images are generated per slice, and are weighted by the diffusion along each corresponding axis.

### 2.4.2.2. The "artifact" of anistropy

In the case of isotropic diffusion, such as that of CSF, we observe an identical signal in all directions of the diffusion gradients:

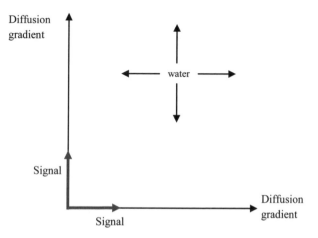

**Figure 2.8.** *Movement of the water molecules and the associated signal, in the case of isotropic diffusion. The length of the arrow, which is proportional to the distance traveled by the molecule in a given period of time, is obviously related to D*

As we saw earlier, a decrease in diffusion in the direction of the diffusive gradient will (in principle! (see section 2.4.2.3)) cause a hypersignal on the image produced from that axis of diffusion.

However, with regard to the nervous system, diffusion hypersignals may appear in a given orientation, even in the absence of any disease-related process: such is the case with the white matter.

The diffusion of water molecules is indeed facilitated in the direction of the axonal fibers, but is reduced in an orientation perpendicular to these fibers.

Thus, a hypersignal may also indicate a perpendicular arrangement of the white-matter fibers in relation to the diffusion gradient: we speak of the *"artifact of anisotropy"*.

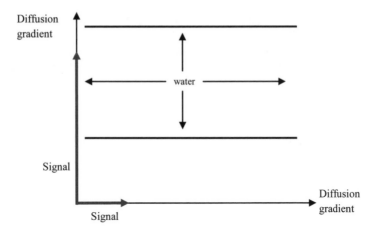

**Figure 2.9.** *Movement of water molecules in a fiber and the associated signal*

### 2.4.2.3. The "artifact" of $T_2$ shine-through

Diffusion is isotropic in the gray matter, and takes place more easily than in the white matter (see the values of D: section 2.3.2):

Therefore, the signal should be less intense.

However, in diffusion images, the gray matter signal is generally hyperintense in comparison to the white matter.

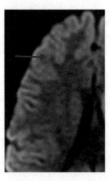

**Figure 2.10.** *Hyperintense signal from the gray matter*

In order to comprehend this phenomenon, it should be noted that the sequence used, the diffusion EPI sequence, gives the signal $T_2$ weighting: this is reflected in the shape of the signal obtained (see section 2.3.1.3).

This is attributable to the fact that the repetition time, TR, chosen is very long (particularly with single shot treatments), which is also true of the echo time TE (between 75 and 100 ms), because this TE needs to include the time necessary for the application of the diffusion gradients.

Structures with a long $T_2$, such as the gray matter, appear hyperintense, identical to those which exhibit reduced diffusion: therefore, there is a risk of confusion when interpreting the image.

This phenomenon is known as "$T_2$ shine-through"; in order to alleviate the confusion it is sometimes necessary to use a different imaging technique; see section 2.4.3.

*Particular case of a newborn baby*

In newborns and young children, the white matter is not yet heavily myelinated, and therefore gives a hyposignal regardless of the direction of the diffusion gradients.

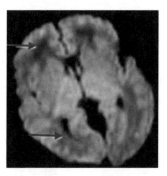

**Figure 2.11.** *Signal from the white matter in a newborn (Reproduced with kind permission from Professeur Catherine Adamsbaum, Radiologie Pédiatrique, Hôpital Bicêtre, APHP, France)*

This effect is reduced with cerebral maturation and progressive myelination: then, in directions perpendicular to the diffusion gradient, we observe a hypersignal comparable to that observed in adults.

### 2.4.2.4. *"Artifact of anistropy" or ischemia*

An in-depth study of an ischemic disorder is given in section 2.5.1; here, though, we shall content ourselves with illustrating possible diagnostic errors arising from a diffusion image.

2.4.2.4.1. The problem

In a diffusion image, if we observe a hypersignal, it is not obvious which conclusion to draw: this may indicate ischemia (and therefore the absence of diffusion) or an artifact of anistropy (no diffusion in the direction of the particular diffusion gradients chosen).

2.4.2.4.2. Solution

One of the ways to deal with this difficulty in diagnostics is to create a fourth image by combining the three previous diffusion images (one for each direction of diffusion gradients), averaging the value of each pixel from these three images.

The image obtained is called an "isotropic image" or a "trace image": it is this image which is usually sent to clinicians.

These "isotropic diffusion" images usually display a relatively homogeneous signal.

However, this technique is not a "miraculous" breakthrough: discrete hypersignals are still often visible along corticospinal fibers, for instance.

The diffusion along these fibers is highly anisotropic. These hypersignals are imperfectly corrected by the averaging of diffusion images acquired in the three orthogonal spatial directions.

2.4.2.4.3. Clinical study

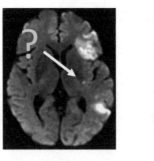

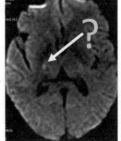

*Case 1*                    *Case 2*

**Figure 2.12.** *Isotropic images (averaging in all three directions) of two clinical cases which are different but which both display a hypersignal*

In clinical cases, we obviously need to distinguish between cerebral ischemia and the artifact of anistropy.

Two additional indicators suggest that a hypersignal represents an artifact:

– bilaterality of the hypersignal;

– the disappearance of the signal if the isotropic image is obtained by averaging over a large number of spatial directions (approx. 25): the image obtained is more homogeneous with the artifact of lessened anisotropy.

In case 2, we note that the hypersignal is vaguely symmetrical.

We therefore suspect it to be an artifact of anisotropy, confirmed by averaging over 25 images: the signal disappears.

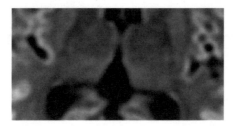

**Figure 2.13.** *Isotropic image obtained by averaging over 25 directions*

The signal was generated by the posterior limb of the internal capsules.

In Case 1, averaging over more than 25 images does not alter the signal: the diagnosis is an acute arterial ischemic disorder.

### 2.4.3. *Apparent Diffusion Coefficient (ADC) maps*

2.4.3.1. *Advantage*

Two major problems are encountered with diffusion images:

– the phenomenon of $T_2$ shine-through, which could lead to difficulty in making a definite diagnosis (see section 2.4.2.3).

– the signal obtained is dependent on the characteristics of the sequence, because its intensity depends on the $b$ factor (see section 2.3.3).

In light of what we have just seen, there is no absolute way of quantifying the diffusion.

Therefore, we shall now turn our attention to seeking an imaging technique and post-processing method whereby we can image only the diffusion coefficient D.

### 2.4.3.2. *Approach*

In order to calculate D, we need to have at least two acquisitions from diffusion imaging:

– One acquisition with no diffusion gradient (b = 0): the signal obtained is then:

$$M_0 = M_T(0)\exp\left(-\frac{T_E}{T_2}\right)$$

– One acquisition including diffusion gradients with a determinate value of b; the signal obtained is then:

$$M_T = M_T(0)\exp(-bD)\exp\left(-\frac{T_E}{T_2}\right)$$

Thus:

$$\frac{M_T}{M_0} = \exp(-b\cdot D)$$

The value of D is then given by the relation:

$$\ln\left(\frac{M_T}{M_0}\right) = -b\cdot D$$

Another method, which is more costly in terms of time, is used to obtain more precise values for D: additional acquisitions are carried out, using different values for b.

The ensemble of the points $\left(b; \ln\left(\frac{M_T}{M_0}\right)\right)$ is represented on a graph: by calculating the slope, we obtain D.

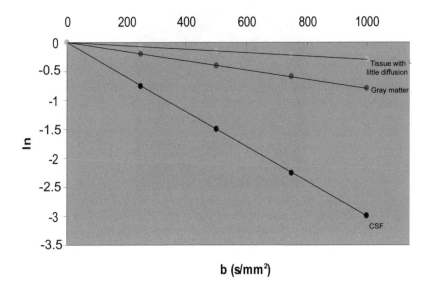

**Figure 2.14.** *Obtaining the diffusion coefficient by varying the value of b*

For each tissue, the greater the slope of the curve, the easier diffusion will take place within that tissue.

Therefore, we can generate a map of diffusion, called an ADC (Apparent Diffusion Coefficient) map, representing, in grayscale for instance, the diffusion coefficients for each pixel. We then speak of the "ADC signal", or quite simply of the "ADC".

*Caution!* On this type of map, the areas with slow diffusion are represented by hyposignals, unlike with diffusion imaging, where they are shown as a hypersignal.

### 2.4.3.3. *An example of a "normal" ADC map*

In this section, we shall demonstrate the advantage of studying diffusion for the treatment of many different accidents or diseases. We shall also analyze the numerous pitfalls that need to be avoided when interpreting the images obtained to reach a correct diagnosis.

Before going into detail about these various problems, it is helpful to have a "reference" image taken from a healthy patient.

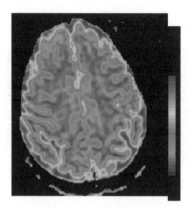

**Figure 2.15.** *ADC map for a healthy patient. For a color version of the figure, see www.iste.co.uk/perrin/MRITech.zip*

As diffusion takes place more easily in the gray matter than in the white matter, the ADC signal is stronger in the gray matter, which therefore appears in red.

The areas shown in blue represent areas where diffusion is slight.

### 2.4.4. *Pitfall images*

*Below, we shall present a number of images which could lead to erroneous diagnoses. These images illustrate the limitations of the credence that we can lend to diffusion images without any other cross-reference.*

2.4.4.1. *Diffusion hypersignal that does not lead to a reduced ADC (the "T2 shine-through" effect)*

The "$T_2$ shine-through effect" involves a diffusion hypersignal from a tissue due to its significant $T_2$.

We have already seen this effect, accounting for the hypersignal generated by the gray matter.

We shall revisit it in our discussion of ischemic accidents (section 2.5.1).

2.4.4.2. *Diffusion hyposignal that does not lead to a strong ADC ("T2 black out")*

The basal nuclei have a high iron content, which gives them a low $T_2$.

Therefore, they exhibit a hyposignal in diffusion imaging: this is the opposite effect to $T_2$ shine-through.

There is a danger of interpreting this as an increased ADC. Therefore, it is necessary to create an ADC map in addition to the diffusion imaging. It should be noted, however, that this effect is not commonly encountered.

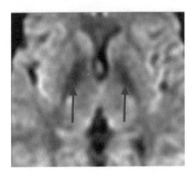

**Figure 2.16.** *Hyposignal from basal nuclei in diffusion imaging*

*2.4.4.3. Normal diffusion that does not lead to a normal ADC ("T2 wash out")*

This is a diffusion isosignal which results from a balance between the $T_2$ shine-through effect and a drop in signal due to the increase in ADC.

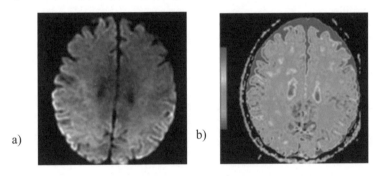

**Figure 2.17.** *(a) Diffusion image; (b) ADC map.*
*For a color version of the figure, see www.iste.co.uk/perrin/MRITech.zip*

The diffusion image (b = 1000 s/mm$^2$) is normal, which could correspond to a normal ADC. In reality, from the ADC map, we can see that the diffusion coefficient in the white matter is increased (compare this map with the one in section 2.4.3.3).

This increased ADC, which decreases the diffusion signal, is compensated by an abnormal elongation of the $T_2$, which increases the diffusion signal: the result is a diffusion isosignal (disease: hypertensive encephalopathy).

## 2.5. The different clinical applications for diffusion

### 2.5.1. *Study of an ischemic accident*

An ischemic accident is characterized by two phases: the emergence of cytotoxic edema and the emergence of vasogenic edema.

2.5.1.1. *Mechanism involved in an ischemic accident*

2.5.1.1.1. Cytotoxic edema

Cytotoxic edema arises within a few hours of a cerebrovascular accident (CVA) and occurs in a number of stages:

1. Occlusion of an artery.

2. Lack of oxygene provision to the cells.

3. Alteration of ion exchanges with the extracellular area:

a. Decrease in ATP production.

b. Breakdown of the sodium–potassium pump whose role is to transport sodium out of the cell.

c. Entry of sodium into the cell.

4. Osmotic entry of water into the cell, causing an increase in volume: this gives rise to cytotoxic edema. The volume of extracellular water is thus reduced.

A recent ischemic accident is characterized by a cytotoxic edema.

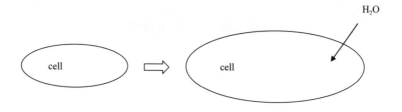

**Figure 2.18.** *Cytotoxic edema: water entry into the cell*

2.5.1.1.2. Vasogenic edema

Next, the osmotic attraction of water from the vascular area to the extracellular area causes the emergence of this new type of edema, which progressively replaces cellular edema.

The volume of extracellular water is increased.

2.5.1.2. *Contribution and limitations of conventional sequences for the detection of an ischemic accident*

2.5.1.2.1. FLAIR sequence and $T_2$-weighted sequences

Water has a very long relaxation time $T_2$.

Hence, $T_2$-weighted sequences are very sensitive means of detecting the increase in the water content of tissues, which occurs during both the main phases of a CVA (cytotoxic edema and vasogenic edema): cytotoxic edema is evidenced by a hypersignal.

Thus, $T_2$-weighted sequences are more sensitive than a scanner to the acute phase of an ischemic accident, because 80% of patients have an abnormal MRI result, whereas only 60% have an abnormal scanner result within the first 24 hours.

The FLAIR sequence eliminates the hypersignal from the CSF by way of the introduction of a specific inversion pulse, and is therefore able to detect the beginning of cytotoxic edema earlier than can $T_2$-weighted sequences. (Detection is rare before the 8th hour with a $T_2$-weighted sequence; with a FLAIR sequence, detection is possible before the 8th hour).

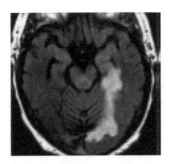

**Figure 2.19.** *Detection of a cytotoxic edema using a $T_2$-weighted sequence. (From [RAD 06]. Reproduced with the kind permission from C. Oppenheim)*

2.5.1.2.2. T1-weighted sequences

A T1-weighted sequence is far less sensitive at an early stage in a CVA than a T2-weighted sequence. Indeed, cytotoxic edema appears later in the form of a hyposignal (in view of the very long T1 of water).

2.5.1.2.3  Problems relating to $T_1$- or $T_2$-weighted sequences in the study of a CVA

The problem with $T_1$ and $T_2$ imaging techniques lies in the fact that the hypersignal in $T_2$ imaging (e.g. with a FLAIR sequence) and the hyposignal in $T_1$ imaging do not change over time.

However, if any surgical intervention is to be performed, we need to know at exactly which point in the progression of the CVA we are intervening: this information is gained by diffusion imaging.

2.5.1.3. *Contribution of diffusion imaging*

2.5.1.3.1. When dealing with cytotoxic edema

In cytotoxic edema, the amount of extracellular water is decreased, but the total amount of water in the tissue increases.

– In diffusion imaging, two mechanisms affect the signal in the same way:

    - the decrease in free movement of water, leading to an increase in signal on the diffusion sequence;

    - the increase in the total amount of water, leading to an increase in the signal on the diffusion sequence by the $T_2$ effect.

Both of these mechanisms cause a *hypersignal*.

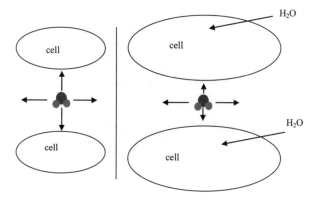

**Figure 2.20.** *Decrease in free movement of water during cytotoxic edema*

– When water is able to move less freely within a given medium, the ADC is decreased: in this case, an ADC map will show a *hyposignal.*

*Shape of a recent ischemia*

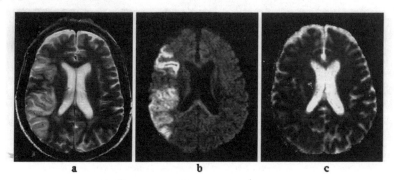

**Figure 2.21.** *Axial slices from a patient who presented a transient cerebral ischemia. With a FLAIR sequence, we see obvious anomalies in the region vascularized by the right middle cerebral artery (MCA); (b) Diffusion imaging with b = 1000: a marked hypersignal in the area of the infarction corresponding to the cytotoxic edema (decreased diffusion); (c) ADC map: hypointense area*

2.5.1.3.2. When dealing with vasogenic edema

During vasogenic edema, the amount of extracellular water is increased, as is the total amount of water in the tissue.

In diffusion imaging, two mechanisms affect the signal, but in the opposite way:

– the increase in free movement of water, leading to a weakened signal on the diffusion sequence;

– the increase in the total amount of water, leading to a stronger signal on the diffusion sequence by the $T_2$ "shine-through" effect.

*Thus, it is impossible to predict the variation in the diffusion signal:* it may be increased, remain normal or be decreased:

– it may prove difficult to evaluate the age of a CVA using diffusion imaging;

– when water is able to move more freely within a given medium, the ADC is increased: therefore the ADC map shows a *hypersignal.*

The properties of areas of diffusion therefore appear equivalent to those of the CSF.

*Shape of an old ischemia*

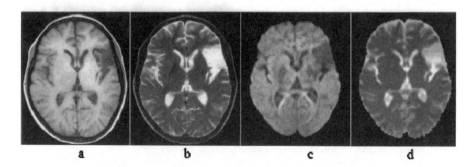

**Figure 2.22.** *Axial slices from a patient presenting an old ischemia in the left middle cerebral artery. The damaged area is hypointense in T₁-weighted imaging (a) and diffusion-weighted imaging (c). The damaged area is hyperintense in T₂-weighted imaging (b) and ADC imaging (d)*

### 2.5.1.3.3. Conclusions

The ADC in an infarction varies over time.

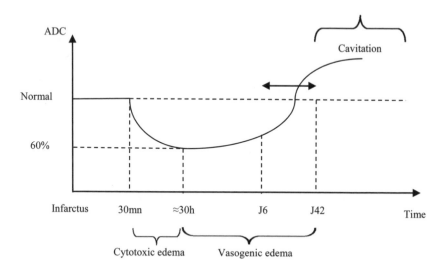

**Figure 2.23.** *Change in the diffusion coefficient over time during an ischemia*

It constitutes a sort of "chronometer" for the ischemia:

1. The ADC is generally normal within 30 minutes after the infarction occurs.

2. After this short period, the ADC decreases progressively under the influence of a cytotoxic edema, lasting anywhere between a few hours and 3-4 days.

3. Later, under the influence of vasogenic edema and cell lysis, the amount of extracellular water gradually increases. This causes the ADC to increase to its normal value. Generally, the ADC is no longer decreased after the 7th-10th day.

4. Finally, the ADC continues to increase because of the phenomenon of cavitation (lost cells replaced by CSF, with a very high ADC).

2.5.1.3.4. Applications: Where is the recent infarction?

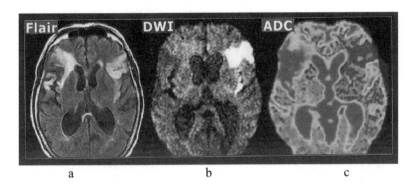

**Figure 2.24** *(a) FLAIR image; (b) diffusion-weighted image; (c) ADC map.*
*(From [SAV03]). For a color version of the figure, see www.iste.co.uk/perrin/MRITech.zip*

This patient has two infarctions which are visible in the FLAIR image.

In view of what we know about the evolution of the ADC in the infarction, can you tell which side the recent infarction is on?

2.5.1.3.5. How can we deal with the problem of $T_2$ shine-through? Other possible techniques

We have seen the advantage of an ADC map for overcoming the $T_2$ weighting of diffusion images. It is also possible to increase the diffusion weighting of images by increasing the b factor.

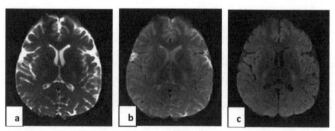

**Figure 2.25.** *Axial cerebral slices using a diffusion-weighted EPI sequence in a patient exhibiting a fungal cerebral infection: (a) image obtained with no diffusion gradient (b = 0); (b) b factor = 500; (c) b factor = 1000*

The vasogenic edema appears as hyperintense in the $T_2$-weighted image (a) but also in the "slightly" diffusion-weighted image (b), illustrating contamination by the $T_2$ shine-through effect.

This is corrected by increasing the b factor of diffusion, making the edema then appear hypointense (c).

The ADC map offers confirmation, showing a hypersignal (d).

### 2.5.2. *Cerebral abscess or necrotic tumor?*

Diffusion imaging and ADC maps can also help to differentiate between an abscess (presence of pus, reduced diffusion) and a necrotic tumor (presence of free water, significant diffusion). Let us take a look at the following two clinical cases:

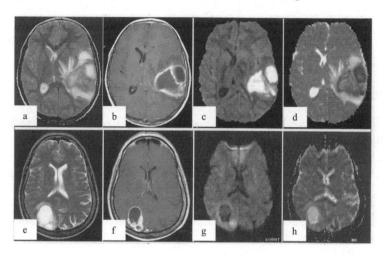

**Figure 2.26.** *Case 1: (a), (b), (c), (d); Case 2: (e), (f), (g), (h)*

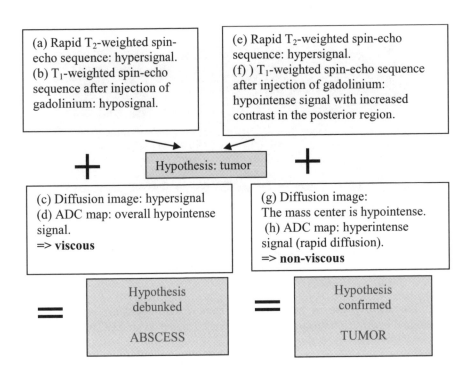

**2.5.3. Limitations of diffusion imaging**

*2.5.3.1. Therapeutic decisions when dealing with a CVA*

We can already see the advantage of using diffusion imaging to give us a timeline for the CVA with which we are dealing. The important thing is being able to tell which area of tissue might yet be able to be saved (called the "penumbral zone"). A different type of imaging will give us this indication: perfusion imaging (see Chapter 3).

*2.5.3.2. The need to fall back on "conventional" sequences*

Caution! Diffusion imaging is not infallible!

It is sometimes necessary to make use of other, more conventional sequences.

*Differentiating a cerebral hemorrhage from an ischemia*

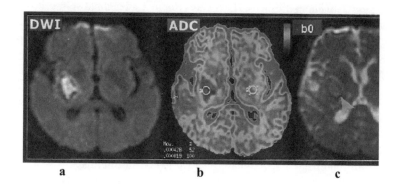

**Figure 2.27.** *(a) Diffusion imaging (b ≠ 0), (b) ADC map,*
*(c) Sequence T₂ (b = 0). (From [SAV03]). For a color version of the figure, see*
*www.iste.co.uk/perrin/MRITech.zip*

In the acute phase, a hematoma (a hemorrhagic accident) is shown as a hypersignal in diffusion imaging with a weakened signal on an ADC map.

This is due to the hyperviscosity of the hematoma.

The characteristics of this signal are therefore similar to those of a recent ischemia (an ischemic accident).

It is obviously of crucial importance to distinguish hematoma from ischemia in order to choose the appropriate course of treatment.

For this purpose, we can use a simple $T_2$-weighted echo-planar sequence: such a sequence may display a hyposignal, often peripheral, in case of a hematoma (remember that we obtain a hypersignal in $T_2$-weighted (or FLAIR) imaging if we are dealing with an infarction).

The hyposignal in the echo-planar sequence is due to hemoglobin degradation products, such as deoxyhemoglobin, which engenders an effect of magnetic susceptibility.

## 2.6. Artifacts frequently encountered in diffusion

*Diffusion sequences are based on the template of an echo-planar sequence: therefore, they include all the artifacts which are known to occur with that sequence. Our aim in this section is not to work on the origin of all these artifacts, but rather to list those which manifest themselves in diffusion images, and gain an understanding of the way in which they might skew the interpretation of the images.*

### 2.6.1. *Artifact specific to EPI*

2.6.1.1. *Foucault currents*

Foucault currents are responsible for image distortion.

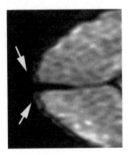

**Figure 2.28.** *Distortion artifact*

This image distortion is problematic in the case of an ADC map or reconstruction if fibers in tractography (see section 2.8).

One way of dealing with this is to use segmentation of the sequence.

2.6.1.2. *Ghosting*

When ghost images are projected onto the image of interest, they may mask a drop in the ADC at that point.

The diffusion image shows the presence of a recent ischemic accident (horizontal purple arrows); this accident is partially masked on the ADC map by the projection of a ghost image (red vertical arrows).

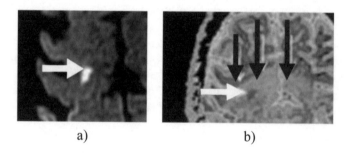

a)                                          b)

**Figure 2.29.** *(a) diffusion image (b = 1000 s/mm²); (b) ADC map.*
*For a color version of the figure, see www.iste.co.uk/perrin/MRITech.zip*

One of the solutions is to increase the FOV (at the cost of a loss in spatial resolution, or an increase in the acquisition time) to remove the ghost image from the image of interest.

### 2.6.1.3. *Chemical dispacement artifact*

This artifact primarily centers on the problem of fat, which can interfere with the measurement of the diffusion coefficient. The problem can be prevented by using a pre-saturation technique.

## 2.6.2. *Artifact peculiar to the diffusion sequence*

### 2.6.2.1. *Nonlinearity of the diffusion gradient*

An error in the estimation of the gradients leads to an error in the estimation of the b factor (see the Stejskal–Tanner formula).

For instance, 5% nonlinearity leads to a 10% error in the estimation of the b factor, and therefore significant errors in the estimation of the diffusion coefficients.

### 2.6.2.2. *Motion artifact*

When taken individually, each diffusion image (for a given direction) is not hugely susceptible to motion artifacts, given the rapidity of the EPI sequence.

Conversely, images such as the trace image, which require several (at least three) separate acquisitions to then be reconstructed, are greatly affected by any movement on the part of the patient. The resulting image may therefore be difficult to interpret.

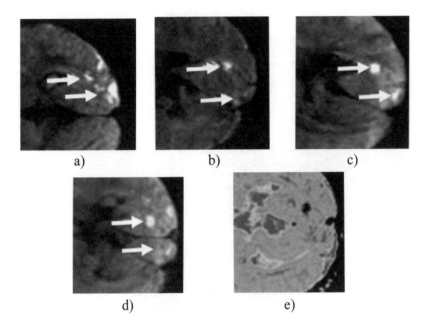

a)                          b)                          c)

d)                          e)

**Figure 2.30.** *(a), (b) and (c): Diffusion images (b = 1000 s/mm$^2$) in the three usual directions (d) Averaged diffusion images (b = 1000 s/mm$^2$); (e) ADC map. For a color version of the figure, see www.iste.co.uk/perrin/MRITech.zip*

The artifacts from the movement of the patient's head (between images (a) and (b)) can clearly be seen in images (d) and (e).

Here, we are seeking to observe an ischemic lesion of the area around the left posterior cerebral artery (horizontal white arrows).

If we notice this type of artifact, it is generally possible to consult the initial images ((a), (b) and (c)), or indeed it is often preferable to begin the acquisition anew (given its rapidity), e.g. after slight sedation.

2.6.2.3. *Artifacts relating to the use of multi-channel antennas*

When using multi-channel surface antennas, the received signal will be stronger near to the antenna than at depth.

In this case, we obtain shine-through from the cortex.

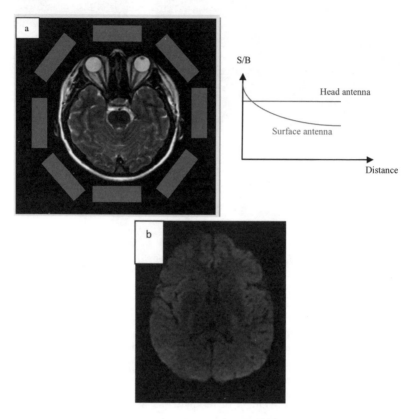

**Figure 2.31.** *(a) Spatial distribution of the antennas and the signal obtained, (b) more intense signal due to the proximity of the antennas*

This artifact again arises in exactly the same way in all the diffusion images, regardless of the value of b. The ADC map therefore does not exhibit this artifact.

## 2.7. Diffusion-tensor imaging

### 2.7.1. *Advantage*

Diffusion imaging such as we have discussed up until now always measures diffusion along an axis that is pre-determined by the operator: if the gradients are applied along a horizontal axis, then the signal will only be sensitive to a horizontal motion.

This is not really problematic if we are measuring the diffusion of free water, which is isotropic (i.e. no stresses are involved). However, if we are measuring the diffusion of water in the brain, for instance, we will often find that one particular axis of diffusion is favored over others (see section 2.1.3).

With this in mind, let us return to the example of ink dripped into a glass filled with water: the "blot" obtained is spherical, which indicates that the movement of the molecules of ink (and of water) is totally random: this is isotropic diffusion – the ink (or the water) diffuses in every direction in the same manner.

The diffusion coefficient is directly related to the size of the sphere, and to determine a sphere completely, we need only one parameter: its radius.

Things become more complicated when the shape assumed by the ink diffusing through the medium becomes oval in 2-dimensional space or an ellipsoid in 3-dimensional space. This is known as anisotropic diffusion, and takes place in biological tissues. *In this case, it becomes impossible to describe this diffusion using only one diffusion coefficient.*

*We therefore need to use a new mathematical tool: the diffusion tensor.*

| | Molecular motion | Vectors describing the motion | Mathematical description |
|---|---|---|---|
| **Isotropic diffusion** | | | D |
| **Anisotropic diffusion** | | | Diffusion tensor |

NOTE.– The arrows in this diagram represent the distance traveled by the molecules in the different directions in a defined time-period $\Delta t$ .

These lengths relate to the diffusion coefficient in each direction: $\Delta x = \sqrt{2D\Delta t}$ .

We have already seen that diffusion in gray matter is isotropic, whereas within the axonal fibers, diffusion is anisotropic. If we are able to quantify this anisotropy of diffusion, then at any point in the brain, we can determine the direction of the fibers running through that point. Thus, indirectly, we are able to deduce a precise image of the axonal fibers making up the brain: this technique is known as *tractography*, and will be discussed in section 2.8.

In order to have a hope of obtaining such results, it is necessary to develop a new type of imaging: *diffusion tensor imaging*.

### 2.7.2. *General principles of diffusion tensor imaging (DTI)*

2.7.2.1. *How can we characterize anisotropic diffusion in 2D?*

If we are able to reconstitute the ellipse seen above, this completely characterizes anisotropic diffusion within the medium.

In order to do so, we understandably need to make measurements of the diffusion coefficients in different directions.

*How many measurements are needed?*

For a more mathematical understanding of this issue, let us take the example of paper, wherein the fibers are oriented vertically.

**Figure 2.32.** *Orientation of fibers in a sheet of paper*

We begin by cutting a disc out of this piece of paper, and rotating it by a certain angle so that we no longer know the horizontal and vertical directions of the paper.

The aim, obviously, is to find these directions.

One of the things we can do is to (again!) drip ink onto our scrap of paper: of course, we expect the stain formed to be an ellipse rather than a sphere, which enables us to identify the vertical directions from the original sheet of paper as being the direction of the longer axis of the ellipse.

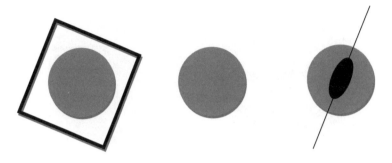

**Figure 2.33.** *Formation of an ink-stain, showing the direction of the fibers*

Now suppose that the ink is invisible, and it is only possible to measure the length of the stain in specific (or arbitrary) directions.

Is it possible to reconstruct the stain by measuring its size only along two axes (x and y)? In order to find out, let us perform two measurements, using two perpendicular axes.

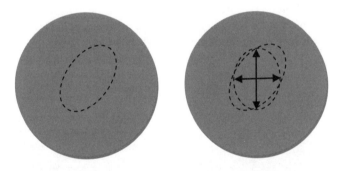

**Figure 2.34.** *Attempt to define the shape of the stain with two measurements*

There is more than one ellipsis which would correspond to these measurements. Hence, we can see that two measurements are not enough to define the shape of the stain. Therefore, we need to increase the number of measurements.

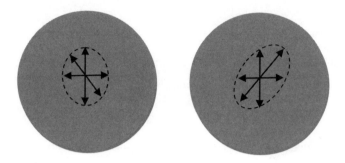

**Figure 2.35.** *Uniqueness of the stain after obtaining three measurements*

As we can see from the two diagrams above, it is in fact necessary to perform three measurements, using three different axes, in order to find the definite shape of the ellipsis.

It is, of course, possible to perform more measurements than necessary to obtain the shape of the stain; if the results obtained were not tainted with error, to do so would lead to a redundancy of information.

Yet in practice, as any measurement we make is liable to contain a certain degree of error (as exemplified by the white double-headed arrow below), a greater number of measurements helps us to better define the shape of the ellipse.

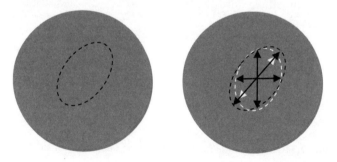

**Figure 2.36.** *Averaging of errors after multiple measurements*

Evidently, this is equivalent to estimating an equation for a straight line: in theory, we only need two points to determine the slope and the ordinate value at the origin; in practical terms, in order to improve the accuracy of the results obtained with these two values, a greater number of measurements are needed.

## 2.7.2.2. *Mathematical modeling*

### 2.7.2.2.1. Eigenvalues and eigenvectors

Our knowledge of the diffusion ellipse's geometry now needs to be modeled by way of a diffusion tensor. Let us look at how this is constructed.

Suppose we have perfect knowledge of the direction and dimension of an axonal fiber.

We begin with a point on that fiber: the fiber has a favored direction of diffusion: $\vec{v_1}$, and a direction in which diffusion is lesser: $\vec{v_2}$.

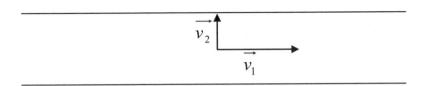

**Figure 2.37.** *Extreme diffusions in an axonal fiber*

The lengths of the arrows shown here are proportional to the diffusion coefficients $\lambda_1$ and $\lambda_2$ in each of the directions (respectively the maximum and minimum in the particular voxel).

If we represent all the diffusion coefficients in each of the spatial directions, we obtain the diffusion ellipse.

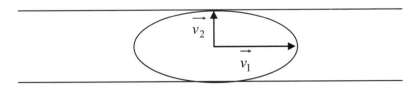

**Figure 2.38.** *Diffusion ellipse*

The directions $\vec{v_1}$ and $\vec{v_2}$ are called "eigendirections" (or main directions) for the diffusion ellipsoid, with which are associated the two "eigenvalues", $\lambda_1$ and $\lambda_2$.

Thus, we create a so-called "diagonalized" diffusion matrix: $\begin{pmatrix} \lambda_1 & 0 \\ 0 & \lambda_2 \end{pmatrix}$.

This matrix perfectly quantifies the diffusion within the medium (the direction, the orientation and the "intensity" of diffusion), as did knowledge of the geometry of the ink-stain in i) above.

2.7.2.2.2. Seeking to find the eigenvectors and their associated diffusion coefficients

Previously, when looking for the shape of the ink-stain, we supposed the ink to be invisible.

In mathematical terms, this is equivalent to not knowing the eigendirections $\vec{v_1}$ and $\vec{v_2}$ or the eigenvalues $\lambda_1$ and $\lambda_2$. They cannot immediately be obtained.

Indeed, the directions of the diffusion gradients used in 2D ($\vec{x}$ and $\vec{y}$) have practically no likelihood at all of actually representing the eigendirections $\vec{v_1}$ and $\vec{v_2}$.

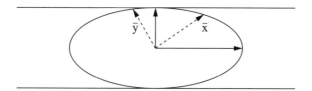

**Figure 2.39.** *Eigendirections and arbitrary directions of the diffusion gradients*

In the base ($\vec{x}$ ; $\vec{y}$), the diffusion matrix becomes: $\begin{pmatrix} D_{xx} & D_{xy} \\ D_{yx} & D_{yy} \end{pmatrix}$

with:

$$\begin{cases} D_2 \leq D_{xx} \leq D_1 \\ D_2 \leq D_{yy} \leq D_1 \end{cases}$$

$D_{xx}$ and $D_{yy}$ are the diffusion coefficients in the directions $\vec{x}$ and $\vec{y}$. $D_{xy} = D_{yx} \neq 0$: the diffusion matrix is no longer diagonal, but it remains symmetrical.

The matrix has three independent coefficients: $D_{xx}$, $D_{yy}$ and $D_{xy}$.

In this situation, we need to perform an operation of "diagonalization" to obtain the eigenvalues and eigenvectors for the diffusion.

As shown by the study in section 2.7.2.2.1., in order to work back to the diagonalized matrix, it is necessary to perform *a minimum of three measurements* of the diffusion coefficients in three independent directions.

### 2.7.3. Diffusion tensor imaging

*In this section, we discuss the general principles set out above, detailing the method of computation and working in the "traditional" 3D space.*

2.7.3.1. Obtaining the diffusion tensor in the base (x, y, z)

*Number of coefficients to be obtained*

In 3D, the diffusion tensor is a positive, symmetrical tensor, which can be represented by a 3×3 matrix: hence, it contains nine coefficients.

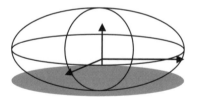

**Figure 2.40.** *Ellipse of the diffusion tensor*

As we can see in this diagram, an ellipsoid can be constructed from three ellipses.

Each ellipsis has one axis in common with the other two.

Finally, it must be remembered that, as we saw in section 2.7.2.1, it is necessary to perform three measurements of diffusion coefficients in three independent directions in order to obtain the shape of each diffusion ellipse.

$$3 \times (3\text{-}1) = 6$$

3 directions

3 independent measurements
for each ellipse

1 direction in common with
the other two ellipses

In order to evaluate the diffusion tensor, it is therefore necessary to carry out six diffusion-weighted acquisitions in six non-collinear directions, as well as the reference acquisition, made in the absence of a diffusion gradient: thus, a minimum of seven measurements are necessary.

*Principle of obtention*

Previously (section 2.4.3.2), in 1 dimension, we defined a signal attenuation equation: $\dfrac{M_T}{M_0} = \exp(-b \cdot D)$ (overlooking the relaxation $T_2$).

In diffusion tensor MRI, the NMR signal attenuation relation becomes:

$$\frac{M_T}{M_0} = \exp\left(-b \cdot g_k^T \overline{D} g_k\right) \qquad [2.7]$$

where    $- g_k \begin{pmatrix} x_k \\ y_k \\ z_k \end{pmatrix}$   is the unitary vector of the direction of the diffusion gradient

$-\ \overline{D}$ is the diffusion tensor: $\begin{pmatrix} D_{xx} & D_{xy} & D_{xz} \\ D_{yx} & D_{yy} & D_{yz} \\ D_{zx} & D_{zy} & D_{zz} \end{pmatrix}$

The value $ADC_k = g_k^T \overline{D} g_k$ is the diffusion coefficient in direction $g_k$ (this is equivalent to a "projection" of the diffusion matrix in the direction $g_k$).

Thus, we can carry out six measurements to obtain six values for the diffusion coefficients $ADC_k$ in six independent directions: $ADC_k = -\dfrac{1}{b}\ln\left(\dfrac{M_T}{M_0}\right)$      [2.8]

The different directions usually sample the space homogeneously. However, the exact coordinates are specific to each constructor.

It is therefore possible to express the different $ADC_k$ values as a function of the elements in the tensor $\overline{D}$:

$$ADC_k = \begin{pmatrix} x_k & y_k & z_k \end{pmatrix} \begin{pmatrix} D_{xx} & D_{xy} & D_{xz} \\ D_{yx} & D_{yy} & D_{yz} \\ D_{zx} & D_{zy} & D_{zz} \end{pmatrix} \begin{pmatrix} x_k \\ y_k \\ z_k \end{pmatrix}$$

Thus, we obtain:

$$ADC_k = x_k^2 D_{xx} + y_k^2 D_{yy} + z_k^2 D_{zz} + 2x_k y_k D_{xy} + 2x_k z_k D_{xz} + 2y_k z_k D_{yz}$$

This gives us six equations, which can be grouped together in the form:

$$\begin{pmatrix} ADC_1 \\ ADC_2 \\ ADC_3 \\ ADC_4 \\ ADC_5 \\ ADC_6 \end{pmatrix} = \overline{X} \cdot \begin{pmatrix} D_{xx} \\ D_{yy} \\ D_{zz} \\ D_{xy} \\ D_{yz} \\ D_{xz} \end{pmatrix} \qquad \Leftrightarrow Y = \overline{X} \cdot \beta$$

The matrix $\overline{X}$ of dimensions (6; 6) depends only on the parameters $x_k, y_k, z_k$ (where $k$ varies between 1 and 6).

The values of the different $ADC_k$ are calculated using equation [2.3], and the coordinates $x_k, y_k, z_k$ are known acquisition parameters. It is therefore possible to work back to all the parameters in the diffusion tensor $\overline{D}$ using the relation: $\beta = \overline{X}^{-1} \cdot Y$.

## 2.7.3.2. *Obtaining the main directions of diffusion*

In order to obtain the main directions of diffusion and the values associated therewith, the tensor $\overline{D}$ needs to be diagonalized.

These directions and values correspond to the eigenvectors and eigenvalues of the tensor and are obtained by solving the following equation:

$$\begin{pmatrix} D_{xx} & D_{xy} & D_{xz} \\ D_{yx} & D_{yy} & D_{yz} \\ D_{zx} & D_{zy} & D_{zz} \end{pmatrix} \begin{pmatrix} x_i \\ y_i \\ z_i \end{pmatrix} = \lambda_i \begin{pmatrix} x_i \\ y_i \\ z_i \end{pmatrix} \Leftrightarrow \overline{D} \cdot v_i = \lambda_i \cdot v_i$$

The eigenvalues $\lambda_i$ are equal to the values of the diffusion coefficients within the framework formed by the vectors $v_i$.

The diagonalized tensor is written as $\begin{pmatrix} \lambda_1 & 0 & 0 \\ 0 & \lambda_2 & 0 \\ 0 & 0 & \lambda_3 \end{pmatrix}$.

## 2.7.3.3. *Graphic representation of the diffusion tensor*

This leads us, finally, back to what we used as our starting point in section 2.7.3.1, when explaining tensor imaging; now we need to visualize the ellipsoid.

We represent it in a framework whose center is the center of gravity of the ellipsoid and whose axes are represented by the eigenvectors.

As we saw previously, if $\lambda_1$ is the diffusion coefficient, the distance of diffusion in direction x over a time-period $\Delta t$ is: $\Delta x = \sqrt{2\lambda_1 \Delta t}$

The characteristic equation of the ellipsoid which we propose to construct is therefore given by:

$$\frac{x^2}{2\lambda_1 \Delta t} + \frac{y^2}{2\lambda_2 \Delta t} + \frac{z^2}{2\lambda_3 \Delta t} = 1$$

This representation enables us to visualize the volume which is covered by the displacement of the water molecules during a given length of time $\Delta t$ of diffusion.

Depending on the type of diffusion, the ellipsoid may assume one of three different characteristic shapes:

– In the case of isotropic diffusion ($\lambda_1 = \lambda_2 = \lambda_3$), the ellipsoid is a sphere.

– When diffusion takes place in a plane ($\lambda_1 \approx \lambda_2 \gg \lambda_3$), we see a "biscuit" shape.

– If diffusion takes place primarily in a particular direction ($\lambda_1 \gg \lambda_2 \approx \lambda_3$), the ellipsoid is greatly elongated in the prevailing direction $v_1$.

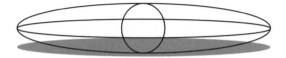

2.7.3.4. *Imaging techniques derived from the diffusion tensor*

The eigenvalues of the tensor $\overline{D}$ can be used to calculate interesting physiological indicators which can be viewed in the form of images.

2.7.3.4.1. Average diffusion

This value is used to obtain an estimation of diffusion within the voxel.

It characterizes the overall displacement of the molecules within the voxel, and can reveal the presence of obstacles to diffusion.

It is calculated by averaging the diagonal coefficients of the diffusion tensor:

$$\langle \lambda \rangle = \frac{D_{xx} + D_{yy} + D_{zz}}{3} = \frac{\lambda_1 + \lambda_2 + \lambda_3}{3}$$

Here are some examples of average diffusion coefficients:

– CSF = $3.19 \times 10^{-3}$ mm$^2$/s;

– gray matter: $0.83 \times 10^{-3}$ mm$^2$/s;

– white matter (corpus callosum): $0.69 \times 10^{-3}$ mm$^2$/s.

2.7.3.4.2. Fractional anisotropy (FA) imaging

There are various indicators that we can use to quantify the anisotropic nature of the diffusion. These too are calculated using the eigenvalues.

Of these, the fraction of anisotropy is the indicator which is most widely used in practice.

$$FA = \frac{\sqrt{3 \cdot \left( (\lambda_1 - \langle \lambda \rangle)^2 + (\lambda_2 - \langle \lambda \rangle)^2 + (\lambda_3 - \langle \lambda \rangle)^2 \right)}}{\sqrt{2 \left( \lambda_1^2 + \lambda_2^2 + \lambda_3^2 \right)}}$$

FA represents the "amplitude" of the diffusion tensor, which can be attributed to its fractional anisotropy. The value of FA ranges from 0 (with isotropic diffusion) to 1 (with infinitely anisotropic diffusion).

FA imaging is commonly used to characterize the anisotropy of diffusion.

2.7.3.4.3. Main diffusion direction imaging

This type of imaging enables us to observe the directions of diffusion in each voxel.

An RGB image is constructed in the following manner:

– Each primary color codes for a particular direction (red: left-right; green: anterior-posterior; blue: cranio-caudal).

– The intensity of each primary color added into a voxel depends on the projection of the principal diffusion vector in the direction corresponding to that color.

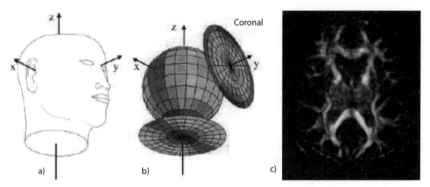

**Figure 2.41.** *(a) Anatomical model; (b) RGB color coding in the main directions (cranio-caudal in blue; left-right in red and anterior-posterior in green); (c) example of an RGB image (axial slice). For a color version of the figure, see www.iste.co.uk/perrin/MRITech.zip*

### 2.7.4. *Clinical applications*

Diffusion tensor imaging is used for all kinds of diseases leading to alteration of the diffusion of water molecules through the tissues, and therefore the structure of those tissues.

#### 2.7.4.1. *Multiple sclerosis*

Multiple sclerosis (MS) primarily affects the white matter, and is characterized by extensive demyelination.

*"Conventional" imaging techniques* are able to detect lesions (or "plaques").

Their primary limitation lies in the lack of correlation between the information provided by these sequences and the evolution of the disease (this is the "clinical/MRI paradox").

*DTI has facilitated the following:*

– in patients suffering from MS, the detection of an increase in radial diffusion (average of the two eigenvalues of the diffusion tensor corresponding to the two directions perpendicular to that of the axon) and therefore an attending decrease in the fractional anisotropy FA;

– the detection, in white and gray matter appearing normal in a "conventional" MRI image, of abnormal alterations of the average diffusion or indeed the anisotropy factor.

Thus, the use of DTI at a very early stage in the progression of the disease could help identify the different mechanisms involved very early on.

It could also guide doctors in making the "correct" choices of treatment depending on the different stages of the disease's development.

### 2.7.4.2. *Tumors*

The choice of therapeutic intervention techniques when faced with a tumor depends mainly on its malignance and grade.

A relation has been observed between the anisotropy factor and the grade of a tumor: the AF of a low-grade tumor is less than that of a high-grade tumor.

Thus, this technique is a non-invasive tool to be used in the making of treatment choices.

## 2.8. Tractography

Tractography is a non-invasive technique which aims to reconstruct the trajectory of the axonal fibers in the white matter on the basis of the data gleaned by diffusion tensor imaging.

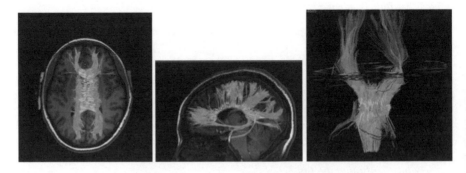

**Figure 2.42.** *Reconstruction of axonal fibers by tractography. For a color version of the figure, see www.iste.co.uk/perrin/MRITech.zip*

There are various methods that can be used to reconstruct the fibers, but here we shall present only one of them, chosen for its simplicity and the regularity with which it is employed by medical personnel: FACT (Fiber Assignment by Continuous Tracking).

## 2.8.1. FACT method

### 2.8.1.1. Introduction

FACT belongs to the family of "deterministic" methods, so called because they determine a unique and definitive direction.

These methods are founded on the idea that the main diffusion direction (given by the eigenvector with the largest eigenvalue) constitutes the tangent to the fiber at that point.

We begin with a chosen source point (or seed point) and propagate a fiber, reconstructed step by step in the main diffusion direction.

FACT goes further still: it takes the main diffusion direction to be that of the trajectory of the fiber itself.

The reconstructed fiber is therefore made up of straight-line segments, whose lengths are equal to the dimension of the voxel in that direction, and therefore vary from voxel to voxel.

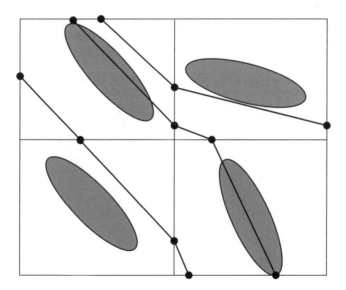

**Figure 2.43.** (a) FACT method for determining the trajectory of a fiber

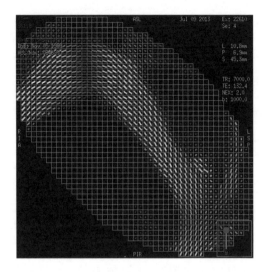

**Figure 2.43.** *(b) Reconstruction at the level of the splenium and the upper forceps*

## 2.8.1.2. *Limitation: precision of fiber assignment*

There are two factors which may skew the accurate reconstruction of the fibers:

– diffusion data are, in general, particularly heavily noised (acquisition noise, physiological noise, etc.);

– the anisotropy fraction (AF) is decreased.

These two factors lead to uncertainty in the determination of the main diffusion vector. Unfortunately, any error at all – even one which is initially relatively slight – can, over the course of repeated iterations, generate a significant difference between the real fiber and the reconstructed fiber. This is called an accumulation error.

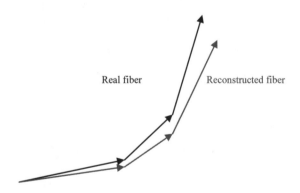

**Figure 2.44.** *Accumulation error in the reconstruction of a fiber*

This error will be all the greater when the step (i.e. the distance between two changes in the direction of the reconstructed fiber) is large.

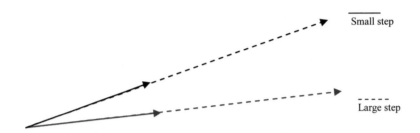

**Figure 2.45.** *Influence of the step on the quality of the reconstruction*

FACT uses steps of varying and relatively significant length (but obviously less than one voxel).

Whilst this method is the simplest and least costly in terms of time, it has poor reproducibility and poor reliability. (In addition to other improvements, other methods also use variable steps, but of shorter length).

### 2.8.2. *Two important parameters in this method*

#### 2.8.2.1. *Choice of seed points*

The starting points for the reconstruction of the fibers, the seed points, are independent of the method of tractography used.

The main difficulty in choosing the "right" seed point is ensuring that the danger of error in the initial direction of the fiber is slight.

For this purpose, we can:

– exploit the anatomical knowledge available to us;

– look for a region containing a group of voxels with a high AF.

The seed point is where the iteration starts: the fibers are reconstituted in the initial, main direction and in both orientations.

2.8.2.2. *Stop criteria*

Axonal fibers cannot propagate indefinitely; two stop criteria are used to stop their reconstruction:

– too low an anisotropy factor (generally less than 0.3):

   - for instance, this is observed in gray matter,

   - in this case, the direction of the eigenvector is too indecisive;

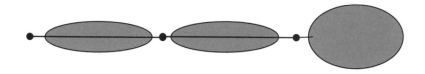

– too radical a change in the orientation of the fiber (which cannot reflect the physiological reality).

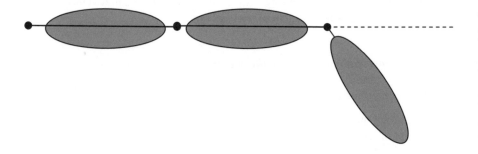

**Figure 2.46.** *Reconstruction stop criteria*

### 2.8.3. *The problem of crossed fibers*

Fiber-crossing is one of the most significant problems in tractography: where this occurs, there is not one but many prevailing diffusion directions.

The tensor model is unable to distinguish between these different directions.

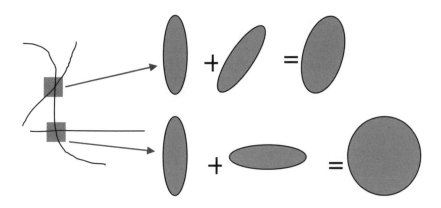

**Figure 2.47.** *Ellipses reconstructed in the case of crossed fibers*

Where fibers cross, two different conclusions may arise:

– the reconstruction of the fiber continues in an incorrect direction;

– the reconstruction of the fiber is stopped because the anisotropy factor is too low.

In any case, neither of these conclusions is satisfactory. Certain techniques, which are more elaborate than FACT, can help to limit errors due to fiber crossing.

### 2.8.4. *Clinical applications*

#### 2.8.4.1. *Multiple sclerosis*

Recently, tractography has made a contribution to the study of this disease.

In the early stages of the disease, demyelination triggers an adaptation reaction: local changes in the trajectory of the axonal fibers, which can be detected by tractography, shows that compensatory "subsystems" are mobilized (by dendritic beading).

#### 2.8.4.2. *Tumors*

The use of tractography may, for instance, help to observe the different alterations and deviations of the axonal fibers due to the presence of a tumor.

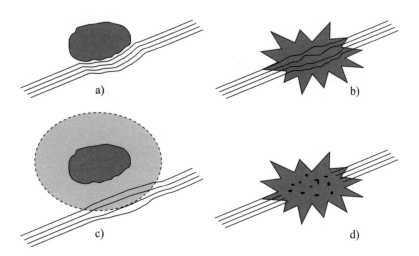

**Figure 2.48.** *Influence of a tumor on the trajectory of the white fibers, (a) diverted trajectory, (b) "infiltrated" fibers, (c) presence of a peritumoral edema, (d) damaged fiber*

Some teams have therefore used tractography for pre-operative preparation.

# Chapter 3

# Perfusion

## 3.1. General points

### 3.1.1. *What is perfusion?*

3.1.1.1. *Differences between angiography and perfusion imaging*

The *macrovascularization* of the encephalon can be studied by magnetic resonance using MRA techniques (see Chapter 1):

– time-of-flight MRA (with or without the injection of a contrast-enhancing product), produces an anatomical image of the main vessels in the encephalon;

– phase-contrast MRA enables us to determine the direction and velocity of the flows.

Cerebral *microcirculation* (in vessels of diameter <200 μm) can also be studied by MRI, using the "perfusion" technique.

The parameters (which will be discussed later on) evaluated by this technique are:

– blood volume;

– time-related data (transit time, time to the contrast peak).

The primary objective of perfusion MRI is therefore to measure (or estimate) the bloodflow irrigating the organ being investigated, i.e. the amount of blood (in milliliters) which passes through 100 g of tissue per unit time (hence, the unit of perfusion is ml/min/100 g of tissue).

NOTE.– This flow corresponds to tissue perfusion (microcirculatory) rather than to the flow along the primary vascular axes.

### 3.1.1.2. *Two different perfusion techniques*

With this technique, we can use tracers which locally modify the signal from the tissue when they enter into a capillary.

These tracers may be:

– *exogenous* (injection of a contrast-enhancing agent).

Exogenous tracers are paramagnetic contrast agents injected intravenously. The use of these contrast agents is at present the most widely used technique in perfusion imaging;

– *endogenous* (no contribution from factors outside the organism itself):

- hemoglobin and deoxyhemoglobin, whose relative concentrations are quantified by the BOLD effect (see Chapter 4),

- arterial protons excited in a particular way before passing into the zone of interest: this is ASL (arterial spin labeling) – we shall discuss the technique in this chapter.

### 3.1.2. *What is the medical advantage to perfusion?*

#### 3.1.2.1. *Advantage in the case of a CVA*

*The maps that can be generated from the temporal data* gleaned by perfusion sequences help to quickly detect hemodynamic problems (abnormal variations in bloodflow); they show the area where an organism's efforts are focused to combat an ischemia.

#### 3.1.2.2. *Advantage in the case of a tumor*

When dealing with a tumor, the *cerebral blood volume map* is of interest: it can be used to determine the grade of the tumor.

*Later on, in section 3.4, we shall give a fuller discussion of the advantage of perfusion techniques in monitoring each of these diseases.*

## 3.2. Exogenous tracers

### 3.2.1. *Technique*

3.2.1.1. *Effect of contrast agent*

The method consists of injecting a gadolinium-based paramagnetic contrast agent. These contrast agents are non-diffusible; they do not cross the hemato-encephalic barrier (they remain in the intravascular space).

The intravenous injection is done rapidly so as to obtain a *bolus of contrast agent*, i.e. a high concentration of the agent in the blood for a short period of time in the cerebral arteries.

*In low concentrations*, gadolinium is used to shorten the relaxation time $T_1$ (in MRA).

*In high concentrations*, gadolinium affects the $T_2$ of the intravascular spins:

– before injection of gadolinium, the magnetic susceptibility is identical in the blood vessels and the surrounding tissues: no alteration of the signal can be seen;

– after injection, when the gadolinium is confined to the capillary microvascularization, it causes a drop in signal by the effect of magnetic susceptibility: the presence of the paramagnetic substance in the vessel causes a local variation in the magnetic field in relation to the surrounding tissues, because of its higher magnetic susceptibility. This accelerates the dephasing of the spins located nearby, reducing their $T_2$ and $T_2^*$, thereby causing a decreased signal in the zone in question, over a distance of a few millimeters around the vessel or vessels.

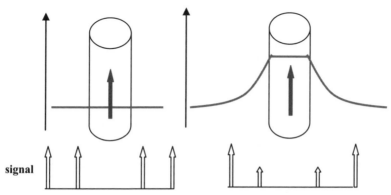

signal

**Figure 3.1.** *Signal in the environs of a blood vessel without (left) and with (right) injection of gadolinium*

The technique of perfusion with injection of a contrast product is known as Dynamic Susceptibility Contrast (DSC).

### 3.2.1.2. *Which sequence should be used?*

Two issues need to be taken into account:

– the effect of injection of the contrast agent is identical to the effect of the magnetic susceptibility artifact; it is therefore most visible with a gradient echo sequence;

– in order to be able to track the change over time in the signal caused by the injection of the contrast product, we need to be able to repeat the acquisition of all of the slices in the region being explored within a very short time-period during that injection. Hence, we have to use an ultra-rapid technique.

For these reasons, we use an *EPI*-type sequence.

### *Characteristics of the EPI sequence*

Coverage of the whole of the brain (in 2-25 slices) in approximately 2 seconds. Sequence repeated during the injection of the contrast product in a bolus (i.e. in high concentration; for this purpose, the flowrate needs to be between 5 and 10 mL/second) for around 45 seconds (approximately 20 phases).

### 3.2.2. *Information obtained with a perfusion sequence*

### 3.2.2.1. *Signal*

We obtain a series of slices (around 20) with a temporal resolution of approximately 2 s.

For each voxel of each slice obtained during the sequence, we obtain a curve $S(t)$ representing the percentage signal loss over time.

### 3.2.2.2. *Which parameters are obtained?*

The curve $S(t)$ has a characteristic form:

– it begins with a rectilinear part, called the baseline, corresponding to a normal signal intensity;

– it dips, when the gadolinium bolus enters the slice, before reaching a peak (maximum concentration of contrast product);

– it recovers progressively until it reaches the level of the baseline once more.

With the doses of gadolinium typically used, the drop in signal is approximately proportional to the concentration of gadolinium.

This curve is then adjusted using a mathematical model: the "gamma function".

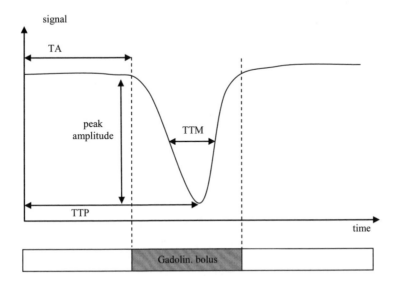

**Figure 3.2.** *Evolution of the signal when gadolinium is injected (at t = 0 s)*

We then define a number of parameters representative of cerebral perfusion in the area under study:

– **TA**: time to arrival of the contrast product in the slice after injection (in s);

– **TTP** (Time To Peak): time corresponding to the maximum variation in contrast (in s);

– **MTT** (mean transit time, in s);

– amplitude of peak: maximum percentage signal intensity loss;

– **rCBV** (regional cerebral blood volume): determined by the area below the curve of signal loss (in mL/100 g of tissue). The rCBV represents the local volume occupied by the blood vessels;

– **CBF** (cerebral blood flow): indicator of blood flow in the brain, corresponding to the ratio of rCBV to TTM.

The signal curves are subjected to post-processing, which enables us to extract the six perfusion parameters, represented in the form of six parametric images (in grayscale or with color coding).

### 3.2.2.3. *Quantification of the parameters*

3.2.2.3.1. Absolute or relative parameters?

The curve representing the evolution of the signal over time with the passage of the contrast product results from two juxtaposed phenomena:

– *passage into the vascular tree:* this is what we are interested in. Thus, we might wish to study, for instance, the velocity at which the blood flow traverses the zone of interest, and use this to deduce the CBF. We use the notation $S_{cap}(t)$ for the signal that would be received if solely the passage into the vascular tree needed to be taken into account;

– *the way in which the contrast agent is injected into the vascular medium:* Obviously, this depends only on the tester. Ideally, the injection should be done in the form of a "bolus": a large quantity delivered instantaneously. In reality, this is of course never the case, particularly because of the injection site some distance from the zone being studied, and because of the volume of product injected. We use the notation $I_{exp}(t)$ for the signal that would be obtained if solely the way in which the gadolinium were injected needed to be taken into account. This assumes that the aspects of capillary networks (which are thin, tortuous, etc.) are eliminated. This function is called the Arterial Input Function (AIF).

In mathematical terms, the signal obtained $S(t)$ can be expressed in the form:

$$S(t) = S_{cap}(t) \otimes I_{exp}(t)$$

The signal obtained $S(t)$ is in fact a convolution between $S_{cap}(t)$ and $I_{exp}(t)$.

Thus, as it currently stands, perfusion imaging is only capable of delivering *relative values of the rCBV or CBF*.

Therefore, very frequently, we look for *ratios* between a diseased area and a health area (which serves as a reference). These ratios are often sufficient to yield an accurate diagnosis (see section 3.4).

3.2.2.3.2.  How can we obtain absolute values?

In order to quantify parameters such as the rCBV or the CBF, we need to obtain the "true" function $S_{cap}(t)$ (which would correspond to that obtained with instantaneous introduction of the contrast product).

With this in mind, it is necessary to perform de-convolution of the curve $S(t)$ obtained with the arterial input function $I_{exp}(t)$.

In practice, $I_{exp}(t)$ is obtained by measuring the intensity of the signal over time in a large artery present in the perfusion image.

*For this purpose, the computer selects a set of voxels which are assumed to be arterial (i.e. which exhibit a curve of short duration with an early peak and a high concentration).*

*The AIF in obtained by finding the mean of the curves in the different voxels. If possible, the pixels are chosen in an arterial area that is close to the damaged area.*

### 3.2.3. *Alteration of the parameters as a function of the type of disease*

The parameters representative of cerebral perfusion vary depending on the types of hemodynamic problems. Therefore, we compare their values in the zone being studied and in a healthy area.

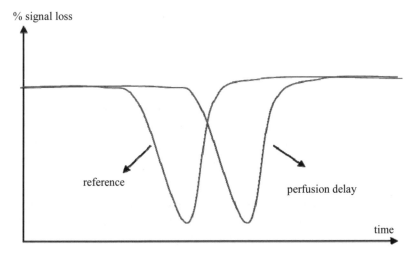

**Figure 3.3.** *Case of a perfusion delay: we observe the shift of the peak decrease in the signal on the timescale*

% signal loss

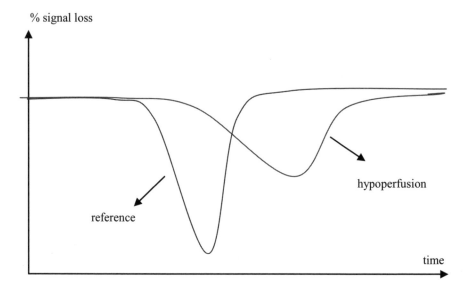

**Figure 3.4.** *Case of hypoperfusion: we observe a decrease in the area below the curve of signal decrease (decrease of the CBV)*

### 3.2.4. *Optimization*

A number of different factors have repercussions for the quality of perfusion imaging:

– the acquisition must take account of only the *first* passage of the bolus into the capillary network (no recirculation);

– the bolus must be injected as quickly as possible: a high concentration of contrast product causes a greater drop in signal strength;

– the hemato-encephalic barrier must not be broken. Indeed, in cases of diseases where the gadolinium escapes from the intravascular medium, there is a reduction in the $T_1$ of the surrounding tissues. This phenomenon acts against the reduction of the perfusion signal $S(t)$; the rCBV (the area below the curve) is therefore underestimated.

– it is crucial to prevent any movement of the patient which would subsequently make it difficult to gather the temporal information relating to a single voxel; the construction of parametric images is no longer possible;

– because of the phenomena of magnetic susceptibility inherent to the EPI sequence, artifacts are visible in perfusion imaging at the boundaries between bone or air and the encephalon.

Thus – similarly to most MRI sequences, but perhaps even more so in this instance – any ferromagnetic material has to be removed (or compensated for by prior spin-echo diffusion imaging sequences).

### 3.3. Endogenous tracers: ASL technique

#### 3.3.1. *Why use endogenous tracers?*

The injection of gadolinium to determine cerebral blood flow (CBF) is used only for measurements of perfusion at a given time. Indeed, because of the persistence and toxicity of the contrast product used, it is difficult to monitor the evolution of the perfusion over time (by rapidly repeating the measurements) during an examination (the half-life of gadolinium is a few tens of minutes).

Hence, we need to find a technique which will deliver a much better temporal resolution.

#### 3.3.2. *Principle and sensitivity of the endogenous tracer method*

3.3.2.1. *Method*

The method involves five steps:

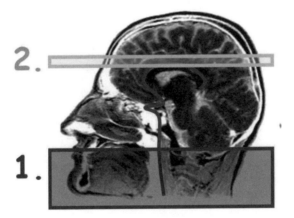

1. A 180° inversion pulse is applied to the water protons present in arterial blood upstream of the slice needing to be imaged. This "labeled" water circulates in the tissues in proportion with the blood flow that they receive.

2. After a time $TI$ needed for the labeled water to reach the slice being imaged, the *labeling acquisition* is performed: the signal then comprises the magnetization in the region of interest, to which is added the magnetization brought in by the protons in the labeled water.

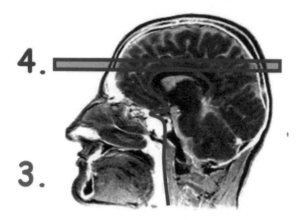

3. The test is then repeated, but this time without labeling the protons in the water in the arterial blood.

4. After the same time period $TI$, the *control acquisition* is performed: the image obtained serves as a reference.

5. The final image is obtained by finding the difference between the labeling and control acquisitions; the signal from the static protons is eliminated, leaving only the signal from the protons in the labeled arterial blood which arrive in each voxel during the time $TI$.

*The signal obtained is therefore directly proportional to the blood flow.*

### 3.3.2.2. Sensitivity

The calculation discussed hereafter (which is easy and quick, because it includes numerous simplifications which we shall discuss later on) is intended to give an order of magnitude of the signal, which can be obtained with the method developed in the previous section.

*In the case of the control acquisition*, the arterial blood has a volumetric magnetization of $M^0$ (magnetization at rest).

If $f$ is the CBF, the magnetization brought into the slice being imaged during the time period $TI$ is $M^0 \cdot f \cdot TI$.

*In the case of the labeling acquisition*, we make the following simplifications:

– we ignore the longitudinal relaxation of the magnetization of the arterial blood;

– we do not take account of the passage of labeled spins of the arterial blood into the tissues (similar to a diffusion process).

Following the application of the inversion pulse, the arterial blood has a volumetric magnetization of $-M^0$.

The magnetization brought into the slice being imaged during the time period $TI$ is related to the CBF: $-M^0 \cdot f \cdot TI$.

By finding the difference between the magnetizations obtained with the labeling and control acquisitions, we obtain $\Delta M = 2M^0 \cdot f \cdot TI$.

Let us estimate that difference in magnetization. In order to do so, we consider:

– an imaged tissue with a blood volume of 100 mL;

– an average blood flow $f$ = 60 mL/mn;

– a time period $TI \approx$ 1s

Thus, $f \cdot TI = 1$mL : the relative variation in magnetization is around 1%.

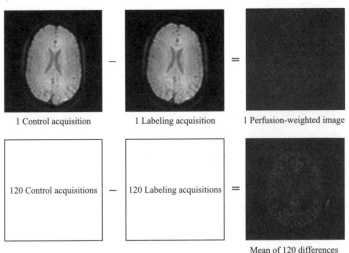

| 1 Control acquisition | 1 Labeling acquisition | 1 Perfusion-weighted image |

| 120 Control acquisitions | 120 Labeling acquisitions | Mean of 120 differences |

**Figure 3.5.** *Influence of repetition of the acquisitions and averaging to obtain a "correct" SNR*

*In conclusion*, from this calculation, we can see that the main advantage of the ASL method is that it yields a signal whose amplitude is directly proportional to the blood flow; its main drawback is that *that amplitude is extremely slight.*

Hence, we can only obtain an exploitable signal by repeating the acquisitions.

The two main techniques of ASL differ in the way in which the labeling pulse is applied. We shall now discuss these two main techniques in detail.

### 3.3.3. *Continuous Arterial Spin Labeling (CASL)*

3.3.3.1. *Sequence*

*In the case of the labeling acquisition*

Continuous arterial spin labeling uses an inversion RF pulse (or a saturation pulse for certain sequences) which is relatively long (1-3 seconds) to continuously label the protons in the water in the arterial blood (for around 1 cm).

For this purpose, a field gradient is applied at the same time as the RF inversion pulse: the resonance frequency of the spins moving in the artery therefore gradually increases. As they pass into the inversion slice (or the labeling zone), the spins are in resonance with the RF pulse: their magnetization is inverted.

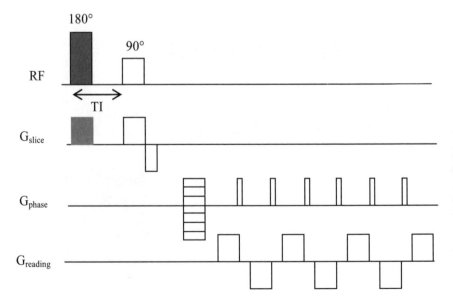

**Figure 3.6.** *Sequence associated with the labeling acquisition*

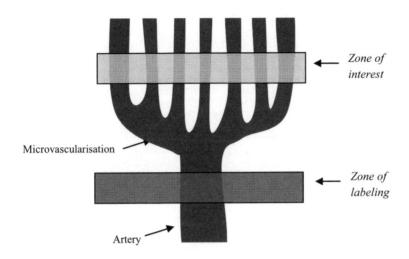

**Figure 3.7.** *Identification of zones of interest and of labeling*

When they arrive in the zone of interest (the slice being imaged), the absolute value of the longitudinal magnetization of the labeled spins has slightly decreased ( $M_z$ is "less negative") because of the relaxation $T_1$, but this amplitude still differs greatly from that of the spins in the tissues of the zone of interest (which have not been exposed to an RF pulse).

Thus, the labeled arterial spins decrease the longitudinal magnetization of the whole of the zone of interest.

In practical terms, after a time period *TI* between the start of the inversion pulse and the start of the 90° excitation pulse, a rapid (EPI) imaging sequence is applied.

*In the case of the control acquisition*

A similar sequence is applied to the labeling acquisition, obviously without the inversion pulse.

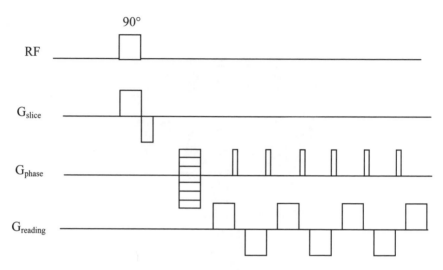

**Figure 3.8.** *Sequence associated with the control acquisition*

**Figure 3.9.** *The control acquisition is made without the inversion pulse*

With the control acquisition done, we need to choose the parameters for the labeling acquisition – particularly the time period *TI*. Such a choice is of crucial importance, because it has a direct impact on the intensity of the signal received.

This time-period depends on a variety of factors, such as the distance $d$ between the labeling zone and the zone of interest, but also the longitudinal relaxation time $T_1$.

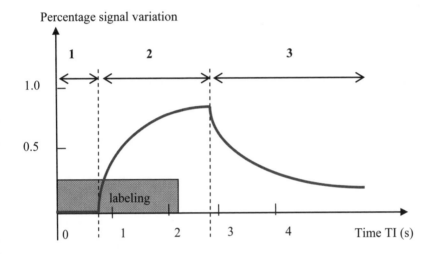

**Figure 3.10.** *Difference in signal (in percentage of the signal from the control acquisition) between the two acquisitions described above, as a function of TI. This curve is obtained for particular values of d and $T_1$*

We can see three main stages in the curve in Figure 3.10:

– in the first part of the curve, the signal is unchanged: the labeled protons have not yet reached the zone of interest;

– the increase in signal in the second part is indicative of the fact that a growing number of labeled protons are entering the zone of interest;

– finally, in the third part, the signal decreases as the labeled protons leave the slice; in addition to this, we see the decrease in magnetization of these labeled spins because of relaxation $T_1$.

### 3.3.3.2. *Principle behind the measuring of cerebral blood flow ("CBF" or "f")*

### 3.3.3.2.1. The equation

We can return to the Bloch equation, taking account of the effects relating to perfusion:

$$\frac{dM_t(t)}{dt} = \frac{M_t^0(t) - M_t(t)}{T_{1t}} + fM_a(t) - fM_v(t)$$

where:

– $f$ is the cerebral perfusion in $mL \cdot g^{-1} \cdot s^{-1}$;

– $M_a(t)$ is the arterial magnetization per mL of blood over time;

– $M_v(t)$ is the venous magnetization per mL of blood over time;

– $M_t(t)$ and $M_t^0$ are the magnetization of the tissues per gram of tissue, over time and at equilibrium;

– $T_{1t}$ is the longitudinal relaxation time of the tissues in seconds.

In the Bloch equation, the terms $fM_a(t)$ and $fM_v(t)$ denote the magnetization of the water protons (present in the blood) which enter (through the arteries) and leave (through the veins) the tissues, as illustrated by Figure 3.11.

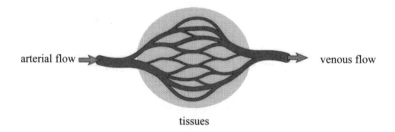

arterial flow ➡                                venous flow

tissues

**Figure 3.11.** *Blood irrigation of tissues*

### 3.3.3.2.2. Simplifying hypotheses

The aim of solving the Bloch equation, which we shall discuss below, is to show the way in which the cerebral blood flow is evaluated with the CASL technique. With this in mind, we are going to use a number of simplifying hypotheses to enable these calculations to be performed relatively easily (if these hypotheses are not made, the calculation is more complex, but the method remains the same).

– In the Bloch equation given in the previous section, we considered that the transit time of the labeled spins into the zone of interest was insufficient for longitudinal relaxation to occur. Hence, the phenomenon does not play a part in this equation (this is tantamount to considering that $T_{1a} \to \infty$). $M_a(t)$ will therefore be taken to be constant throughout this study.

– At equilibrium, the difference in the concentration of water in the blood and in the tissues (and therefore the difference in their magnetization) brings into play the water partition coefficient $\lambda$ (in $mL \cdot g^{-1}$): $M_a^0 = M_v^0 = \dfrac{M_t^0}{\lambda}$, with $M_a^0$ and $M_v^0$ representing the arterial and venous magnetizations per mL of blood at equilibrium.

– At all times, we can make the hypothesis of a complete exchange between the water from the blood which has traversed the tissue and the water in the tissue, so the venous magnetization can be considered to be in constant equilibrium with the tissue magnetization, and we can write:

$$M_v(t) = \frac{M_t(t)}{\lambda}$$

– $M_a(t)$ being constant: $M_a(t) = M_a^0 = -\dfrac{M_t^0}{\lambda}$.

### 3.3.3.2.3. Solution of the Bloch equation and discussion of the method

An inversion pulse is applied to the arterial water at time $t = 0$: $M_a(t = 0) = -M_a^0$.

With the application of the simplifying hypotheses seen above, we obtain the solution:

$$M_t(t) = \frac{M_t^0}{1 + \dfrac{fT_{1t}}{\lambda}} \times \left[ \left( 1 - \frac{fT_{1t}}{\lambda} \right) + \frac{2fT_{1t}}{\lambda} \exp\left( -t\left( \frac{1}{T_{1t}} + \frac{f}{\lambda} \right) \right) \right]$$

Spin labeling is done continuously (labeling pulse with duration $T_{RF}$ between 2 and 3 seconds: thus, we consider that $T_{1t} \ll T_{RF}$).

We then soon reach a steady state (ss): we adopt the expression of $M_t(t)$ as $t \to \infty$ : $M_t^{ss} = M_t(\infty)$

Thus, we obtain: $\dfrac{M_t^{ss}}{M_t^0} = \dfrac{1 - \dfrac{fT_{1t}}{\lambda}}{1 + \dfrac{fT_{1t}}{\lambda}}$

The perfusion can then be expressed by:

$$f = \frac{\lambda}{T_{1t}} \left( \frac{M_t^0 - M_t^{ss}}{M_t^0 + M_t^{ss}} \right).$$

However, if we return to the expression of $M_t(t)$ above, we note that it is possible to define an apparent longitudinal relaxation time: $\dfrac{1}{T_{1app}} = \dfrac{1}{T_{1t}} + \dfrac{f}{\lambda}$.

Thus, we obtain:

$$f = \frac{\lambda}{T_{1app}} \left( \frac{M_t^0 - M_t^{ss}}{2M_t^0} \right)$$

*In order to quantify CBF, therefore, we need to measure the $T_{1app}$ and the magnetizations $M_t^0$ and $M_t^{ss}$*

NOTE.– This calculation does not take account of the effects of the variable transit time of the labeled spins (this is the problem of quantification, which will be discussed later on in section 3.3.6.2).

3.3.3.3. *Problems relating to this technique*

3.3.3.3.1. Problem of magnetization transfer

With ASL, the difference between the signals from the control- and marking acquisitions is around 1%. It is clearly important that all sources of differences be connected to differences in perfusion rather than other parasitic phenomena.

*Therefore, two important points need to be verified:*

– the spins need to be relaxed before we can begin the MRI sequence;

– the static spins must generate the same signal in the control image and the labeled image (necessary for perfect subtraction of the signal).

This latter point is not respected, as we shall see.

The long pulse used for the inversion and centered on the labeling zone is off-resonance for the zone of interest and should usually have no effect on it.

This is indeed what happens with *"free" protons* (those in fat or fluid): they are "too distant in terms of frequency", as shown by Figure 3.12.

On the other hand, the pulse does have an effect on the *"bound" protons*, i.e. those protons in the macromolecules of the tissues in the zone of interest (because their frequency spectrum is very broad). By magnetization transfer, the free protons in the zone of interest are, in turn, affected.

The signal from the free protons in the zone of interest is all the more affected when the *duration of the inversion pulse is long, which is the case with CASL.*

NOTE.– Here, we draw a distinction between free spins and bound spins, because it is primarily the free spins that will be affected by the excitation pulse in the control acquisition, given their narrow frequency range. On the other hand, only a small proportion of bound spins will be affected by the pulse, because of their extensive frequency spectrum.

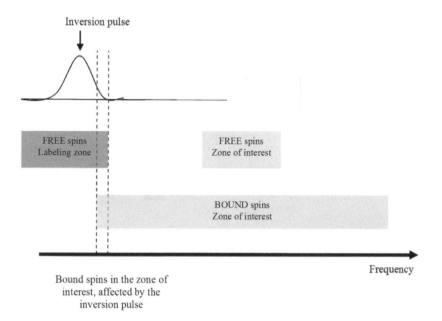

**Figure 3.12.** *Proportion of free and bound spins in the labeling zone and the bound zone affected by the inversion pulse of the labeling acquisition*

This magnetization transfer has effects for the evaluation of the perfusion, as shown by Figure 3.13.

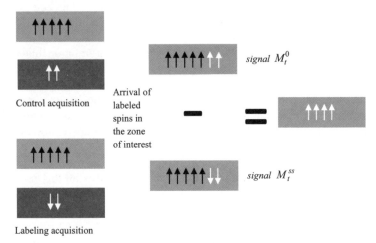

**Figure 3.13.** *Obtaining the signal difference between the control- and labeling acquisitions in the "ideal" scenario – i.e. where there is no magnetization transfer (dark gray background: labeling zone; light gray background: zone of interest. Black arrows: static spins; white arrows: free spins in the labeling zone affected by the inversion pulse)*

NOTE.– The real proportion of static spins in relation to the spins in the blood flow is not respected (see section 3.3.2.2).

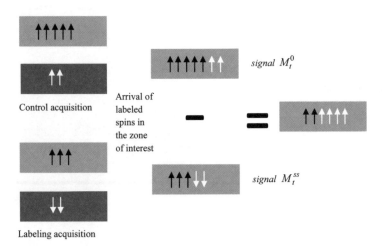

**Figure 3.14.** *Obtaining the signal difference between the control- and labeling acquisitions in the presence of magnetization transfer (legend same as for Figure 3.13)*

*Conclusion*: the magnetization transfer thus leads to the obtainment of a stronger signal, which results in the *overevaluation of the perfusion*.

*In order to solve this problem*, we need to reproduce these effects of magnetization transfer in the control acquisition, by also including an inversion pulse in that acquisition.

The inversion pulse in the control acquisition must be off-resonance in relation to the zone of interest, with the same frequency shift that occurs with the inversion pulse in the labeling acquisition. (In other words, the protons in the macromolecules in the zone of interest are affected in the same way during both the control- and labeling sequences).

Thus, with the CASL technique, an inversion pulse is also applied in the control image, but on the other side of zone of interest, so as to avoid re-marking the spins in arterial blood entering into the slice being imaged.

NOTE.– Often the inversion zone used in the control image may even be outside the patient's head, which means that no labeling really takes place (see Figure 3.15 below).

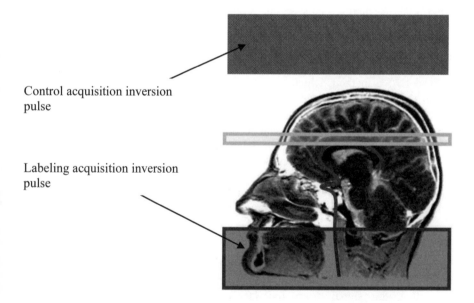

**Figure 3.15.** *Position of the inversion pulse in the control acquisition in order to reproduce the effects of magnetization transfer that occur with the inversion pulse in the labeling acquisition*

NOTE.– The problem is only solved when only one zone (or slice) is being imaged (this zone needs to be placed at an equal distance from the two inversion pulses). Thus, in the case of a *multi-slice acquisition*, this solution is no longer viable.

*How are we to solve the problem of magnetization transfer in the case of a multi-slice acquisition?*

A method has been developed which uses two RF antennas. A surface antenna, placed on the carotids, labels the arterial spins without causing magnetization transfer, and the second antenna is used for the acquisition. The control image can thus be acquired without a pulse. The problem of a multi-slice acquisition is thereby resolved.

Another advantage of this technique is that by perfoming labeling on the arteries with small antennas, we can also measure the perfusion in a vascular area by labeling one artery specifically.

Technically speaking, this solution poses problems: this is one of the reasons why CASL has gradually been abandoned, in favor of its cousin, the PASL technique.

NOTE.– As we shall see later on, the problems of magnetization transfer also affect PASL sequences, but to a lesser extent.

3.3.3.3.2. Problem of SAR

The CASL method, which uses long RF pulses for the labeling- and control acquisitions, leaves a great deal of energy in the tissues (SAR: specific absorption rate), which limits the usefulness of this technique in humans, because these pulses are transmitted by an RF antenna which contains all of the head.

This problem is all the more prevalent when strong fields are used.

Here, once more, we can use surface antennas to decrease the SAR in comparison to techniques using a voluminous antenna for the labeling.

3.3.3.3.3. Problem of transit time

The transit time is the time taken by the labeled blood to reach the slice being imaged (i.e. the zone of interest). When dealing with certain diseases, this time is so long that the signal from the labeled blood has attenuated significantly ($T_1$ relaxation) before it reaches the zone of interest. For simplicity's sake, this relaxation was not taken into account in the workings in section 3.3.3.2.

Generally speaking, the transit time is between a few hundred ms and 1 s. This problem is partially resolved if we work with a strong field (the relaxation time $T_1$ is prolonged when $B_o$ is increased).

Shorter transit times strengthen the signal, but prevent good quantification of perfusion (see the issues of heterogeneity of transit time discussion in section 3.3.6.2).

### 3.3.4. PASL (Pulsed Arterial Spin Labeling) technique

Pulsed arterial spin labeling uses a short RF pulse (of a few ms) to label a large volume of arterial blood (>10 cm).

#### 3.3.4.1. Principle

The major problem with CASL is the deposition of a substantial amount of energy in the patient's body. PASL techniques were developed primarily to deal with this issue.

The various PASL sequences are essentially constructed in the same way as the CASL sequence, but this time the pulses used are *short*.

Thus, exactly as happens with the CASL technique, during the control acquisition a volume of blood is labeled, upstream of the region of interest, by a 180° RF pulse, short so as to limit magnetization transfers.

In both the control acquisition and the labeling acquisition, the signal is received after a time-period *TI* (between 1 and 2 s).

The difference in signal between the measurements with and without spin labeling reflects the amount of labeled blood which has reached the volume of interest in the course of *TI*, which enables us to estimate the cerebral blood flow.

The peculiarity of this method relates to the *limited lifespan of the endogenous tracer*, because the magnetization of the labeled blood bolus will return to equilibrium (depending on the time constant $T_1$ of the blood). This limits the duration of the interval between labeling and acquisition of the signal.

#### 3.3.4.2. Problems with the technique

3.3.4.2.1. Imprecision of the inversion pulses

With PASL, the spatial selectivity of the inversion pulses is poor. Owing to this "imperfect" selectivity, the labeling pulse alters the magnetization of the static spins in the zone of interest which are closest to the labeling zone (this is known as contamination).

*To eliminate this problem*, we can:

– increase the distance between the zone of interest and the labeling zone. If we do that, we need to be careful to take account of the endogenous tracer's limited lifespan, and therefore not increase the distance by too much (otherwise there would no longer be any difference between the labeling- and control signals);

– saturate the spins in the zone of interest during both the control acquisition and labeling acquisition. This solution offers a twofold advantage:

- it lessens the contamination of the inversion pulse,

- it eliminates the signal from the static spins after subtraction of the two (control and labeling) images.

### 3.3.4.2.2. Magnetization transfer

As with CASL, magnetization transfer can be a source of artifacts which we need to try to eliminate.

However, with PASL techniques, this problem is less prevalent, given the short duration of the pulses used.

*In order to eliminate this problem*, we can use the same approach as with the CASL technique, adding to the control acquisition an inversion pulse situated opposite the labeling pulse in relation to the region of interest (the effects of magnetization transfer will then be identical in both acquisitions).

### 3.3.4.3. *The various PASL sequences*

There are a number of sequences that are widely used; here, we shall discuss only the three most commonly employed.

### 3.3.4.3.1. The "basic" sequence: EPISTAR

The first pulsed labeling sequence is known as EPISTAR (Echo Planar Imaging and Signal Targeting with Alternating Radiofrequency).

This sequence comprises:

– For the labeling acquisition: a *selective* spatial inversion pulse applied to the arterial blood spins (labeling zone).

– For the control acquisition: an inversion pulse situated on the opposite side of the zone of interest from the inversion pulse in the labeling acquisition (meaning that the magnetization transfers are the same for both acquisitions).

– A pulse to saturate the zone of interest (this prevents contamination by the inversion pulse).

 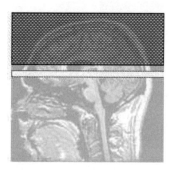

**Figure 3.16.** *Position of the areas affected by the inversion pulse (gray hatched shading) in the labeling acquisition (left) and in the control acquisition (right) in relation to the zone of interest (white bar) in the EPISTAR sequence*

*This sequence is very similar to that discussed when describing the CASL technique in an attempt to deal with the problem of magnetization transfer. The only alterations here are the length of the inversion pulse (shorter with PASL) and the saturation pulse.*

*Drawback to this sequence*

Compensation for magnetization transfer (achieved by virtue of the inversion pulse used in the control acquisition) is only possible for a single target zone, because the effects only cancel one another out in a particular position.

With this sequence, therefore, it is not possible to deal with magnetization transfer if making a multi-slice acquisition.

3.3.4.3.2. FAIR sequence

This sequence comprises:

– For the labeling acquisition: a *selective* spatial inversion pulse applied to the spins in the *zone of interest*.

– For the control acquisition: a *non-selective* inversion pulse.

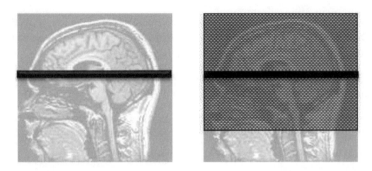

**Figure 3.17.** *Position of the areas affected by the inversion pulse (gray hatched shading) in the labeling acquisition (left) and in the control acquisition (right) in the FAIR sequence*

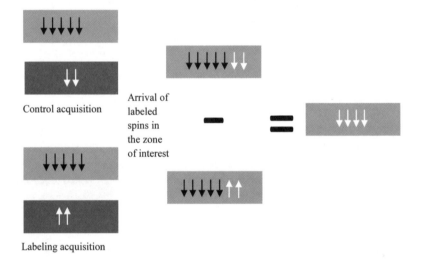

**Figure 3.18.** *Obtaining the signal difference between the control- and labeling acquisitions in the case of the FLAIR sequence (legend same as for Figure 3.13)*

*Advantages of this sequence*

As the same inversion pulses (with and without a selection gradient respectively for the control acquisition and the labeling acquisition) are used in both acquisitions, *the effects of magnetization transfer cancel one another out.*

In addition, the position of the inversion pulse is such that *there is no contamination by the inversion pulse* (as can be caused by the problems of selectivity of this pulse, as we saw in the case of the EPISTAR sequence): hence, there is no need for a saturation pulse.

*Drawback to this sequence*

The non-selectivity of the inversion pulse during the control acquisition means that the venous blood flowing in from the slice above the zone of interest is also labeled: both the arterial blood and venous blood are imaged simultaneously.

3.3.4.3.3. PICORE sequence

This sequence comprises:

– For the labeling acquisition: a *selective* spatial inversion pulse applied to the spins in the *labeling zone*.

– For the control acquisition: a *non-selective* inversion pulse (because no gradient is applied) with the same *frequency offset* as the labeling pulse in relation to the zone of interest.

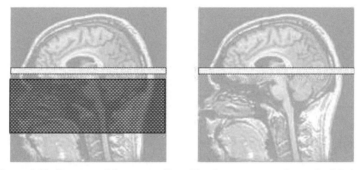

**Figure 3.19.** *Position of the areas affected by the inversion pulse in the labeling acquisition (left) and in the control acquisition (right) in the PICORE sequence*

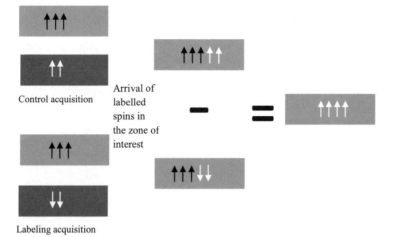

**Figure 3.20.** *Obtaining the signal difference between the control- and labeling acquisitions in the case of the PICORE sequence (legend same as for Figure 3.13)*

*Advantages to this sequence:*

During the control acquisition, the inversion pulse is off-resonance for all free spins. On the other hand, this pulse does indeed have an effect on some of the spins in the macromolecules, particularly those in the zone of interest (see Figure 3.21).

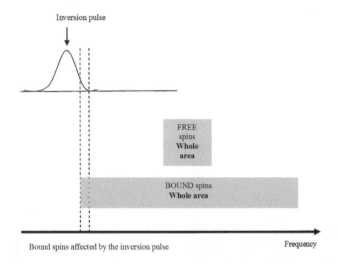

**Figure 3.21.** *Spins affected by the inversion pulse during the control acquisition in the PICORE sequence*

Hence, we see magnetization transfer from the spins in the macromolecules to the free spins in that zone of interest.

This magnetization transfer is identical to that which takes place during the labeling acquisition.

Thus, by the operation of subtraction, *these two magnetization transfers are eliminated.*

In comparison to the FAIR sequence, in this case there is no labeling of the venous flow, and the main advantage over the EPISTAR sequence (namely the possibility of taking multi-slice acquisitions) is preserved.

### 3.3.5. *CASL or PASL?*

The choice of a sequence belonging to one or other of these two major categories depends on a number of criteria, which are listed below:

*CASL method*

*Advantages:*

– no contamination by the inversion pulse, because the labeling zone is slender (long pulse);

– better SNR than with PASL.

*Disadvantages*:

– effects of off-resonance (magnetization transfer) because the pulse is long, and the spins in the macromolecules "have the time" to become saturated;

– problem of SAR.

*PASL method*

*Advantages:*

– Magnetization transfer reduced because of the shorter pulse.

*Disadvantages:*

– Imprecision of the RF pulse, which "spills over" into the zone of interest.

### 3.3.6. Conclusion: advantages and limitations of ASL

3.3.6.1. *Advantages of the ASL technique*

We can hold up two main qualities of ASL:

– non-invasive technique;

– technique which can be repeated an infinite number of times, and which enables us to study the microvascularization and its changes over time.

3.3.6.2. *Its limitations*

The various problems encountered with each of the methods (PASL and CASL) can be resolved by way of the different solutions put forward.

Nevertheless, there are still certain problems that we face when using ASL techniques.

*i) Problem of quantification of the method*

Numerous problems stand in the way of finding a simple relation between the signal obtained and the perfusion value:

– the transit time of the blood is variable in spatial terms: in Figure 3.22, we note that the signal received in the zone of interest varies on the basis of the moment when the acquisition is made;

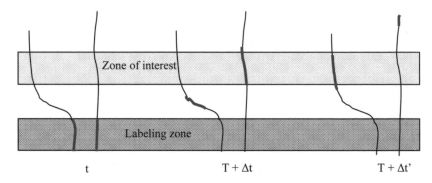

**Figure 3.22.** *Arrival of the labeled spins in the zone of interest at different times depending on the microcapillary path that is taken*

– relaxation of the magnetization labeled during the test over the course of the transit time;

– the specific variations of $T_1$ in space.

A number of modifications to ASL sequences have been made (such as QUIPSS and QUIPSS II) to deal with these problems by applying additional saturation pulses to the zone of interest or labeling zone.

NOTE.– *All perfusion techniques* (both exogenous and endogenous) are afflicted by this problem of quantification of flow, but the origins of such problems are different:

– with an exogenous technique, the problem relates to our knowledge of, or lack thereof, the arterial input function;

– with an endogenous technique, the problem stems from the spatial dependence of the blood's transit time.

*ii) Acquisition with a poor SNR*

In this case, we need to carry out a substantial number of scans, which means the method becomes lengthy to implement.

*iii) Movements* which cause errors in the subtraction of the images (this is why we use echo-planar-type sequences).

*iv) Physiological variations*

These variations, which are non-pathological in origin, can be seen in the microvascularization from one individual to another. They *render it difficult to compare measurements* taken using the ASL technique.

*v) ASL does not work if there is a significant delay in the perfusion*, as happens with a stenosis (because, in this case, the relaxation of the labeled spins before reaching the zone of interest causes the absence of a signal).

*vi) Impossibility of examining the encephalon in its entirety.*

## 3.4. Medical applications

Owing to the possibility of exploring the macrovascularization which they offer, perfusion sequences exhibit an advantage when studying:

– vascular diseases: particularly CVA;

– tumoral diseases: by analyzing the vascularization of a tumor, we can guide diagnosis and therapeutic monitoring;

– infectious or inflammatory diseases.

### 3.4.1. *Ischemic CVA*

3.4.1.1. *Parameters of perfusion imaging and CVA*

An area suffering an ischemia is identified by (see section 3.2.3):

– a decrease in the area under the curve in the perfusion signal (decreased CBV and CBF);

– a shift in the peak decrease in that signal (a greater TA).

The temporal parameters (such as the TA, TTP and TTM) are the most widely used because their respective values are almost identical in the white and gray matter: a hemodynamic problem, caused by an AVC, is therefore very easily identifiable on a map of one of these temporal parameters.

The parameters CBV and CBF are less frequently used, as their values are significantly different between the two forms of matter (meaning that it is less easy to see any anomalies on a map).

3.4.1.2. *Combined contribution of perfusion- and diffusion-based image techniques in treating a CVA*

3.4.1.2.1. Mismatch zone

Emergency treatment when dealing with an ischemia (resulting from an arterial obstruction) is to re-establish the flow: this is what therapeutic attempts at thrombolysis aim to do.

Yet the risk with thrombolysis is hemorrhagic complication, which is often extremely serious, which tends to occur in particular in cases of secondary re-perfusion (without medical intervention) of the ischemiated area.

Perfusion imaging can thus guide therapeutic decisions if used to examine what we call the "mismatch zone".

This zone represents the difference in volume between:

– the zone suffering an ischemia and which is detected by diffusion imaging (reduced diffusion generates a hypersignal);

– the zone of reduced perfusion.

This zone, also called the "penumbral zone", corresponds to tissue which is hypoperfused (the zone of "ischemic sufferance"), but not yet necrotic, and which is likely to be recoverable with thrombolysis treatment.

3.4.1.2.2. Clinical case study

The case discussed below is that of a patient suffering from an acute ischemic accident with aphasia (loss of linguistic function).

In the first acquisitions, taken two hours after the accident, we can see (Figure 3.23):

– a blockage of the proximal segment of left sylvian artery, visible in the MRA sequence;

– anomalies of perfusion which delimit the damaged area: in this area, the temporal factors TA and TTM are much greater than in the healthy area. The damaged area thus defined is far more extensive than the lesion visible with diffusion imaging, which shows an extensive mismatch zone (in red on the TA map).

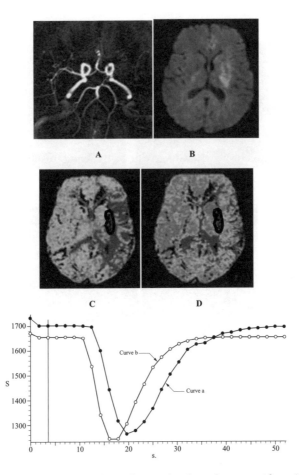

**Figure 3.23.** *Acquisitions made two hours after the ischemic accident. A: MRA;*
*B: Diffusion imaging; C: TA map; D: TTM map; E: Perfusion signal*
*(curve a: damaged area; curve b: healthy area)(Adapted from [GRA 07]). For a color*
*version of the figure, see www.iste.co.uk/perrin/MRITech.zip*

The control acquisitions, made three days later, *after thrombolysis*, show:

– a profound lesion in the sylvian area, visible on the diffusion image;

– a near-normalization everywhere else, shown by the different maps and perfusion curves;

– the disappearance of the occlusion in the left sylvian artery (which correlates to the end of the aphasic episode.

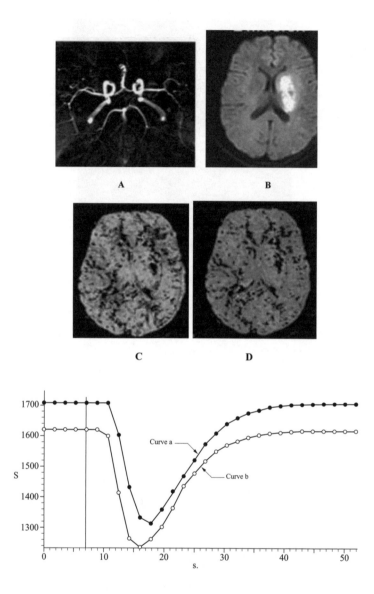

**Figure 3.24.** *Acquisitions made three days after thrombolysis. A: MRA;
B: Diffusion imaging; C: TA map; D: TTM map; E: perfusion signal
(curve a: damaged area; curve b: healthy area)(Adapted from [GRA 07]).
For a color version of the figure, see www.iste.co.uk/perrin/MRITech.zip*

3.4.1.2.3. Conclusion

Thrombolysis will be carried out if both of the following observations present themselves:

– presence of an anomaly of perfusion which is more extensive than the anomaly of diffusion (2/3 of cases encountered);

– presence of a vascular occlusion on an MRA image.

The issue remains open for debate if:

– the areas of anomalies of diffusion are too extensive;

– there is the presence of a mismatch without a vascular occlusion;

– there is the presence of an occlusion without a mismatch.

### 3.4.2. Tumors

3.4.2.1. *Development and grades of a tumor*

Tumors present four stages, or grades, of development (grade I: tumor curable by surgery, to grade IV: average survival six months to two years).

Tumoral development exhibits two distinct phases:

– an initial phase, when the volume of the tumor remains stable: this is the pre-vascular phase;

– a second phase when the tumor needs an additional supply of oxygen and nutrients: this is the vascular phase;

During this phase, we see the formation of new blood vessels from the pre-existing vessels: this is called angiogenesis. The blood volume around the tumor is increased.

Angiogenesis is directly linked to the grade of the tumors, which is obviously helpful for vital prognostication.

3.4.2.2. *Perfusion and support for prognostication or intervention*

When studying a cerebral tumoral disease, of all the parameters that are available to us in perfusion imaging, only the cerebral blood volume is of interest to us for examining angiogenesis.

Technically, ROIs are positioned in a diseased zone and a healthy zone (serving as a reference), as shown by Figure 3.25.

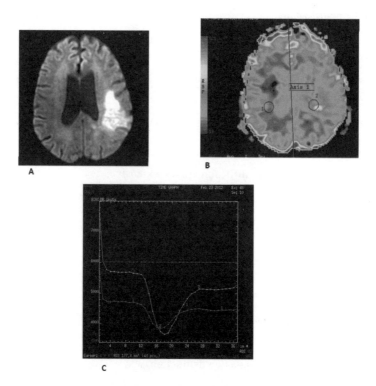

**Figure 3.25.** *Examination of a glioma: A: Diffusion (b = 1000); B: NEI (Negative Enhancement Integral) map (values proportional to the regional cerebral blood volume (rCBV)) and marking of ROIs; C: Shape of signal for each of the ROIs under study*

The quantitative results are presented in the form of a ratio.

Thus, the ratio $rCBV = rCBV_{tumoral\ area}/rCBV_{healthy\ area}$ can be used to characterize the tumor. For instance, in high-grade gliomas, this ratio is generally greater than 2:1; in hypervascular metastases (kidneys, melanoma), it is higher still, etc.

A map of this ratio enables us to demonstrate the differences in vascularization in the zones surrounding the tumor, and can therefore help to guide a surgical intervention (e.g. a biopsy, etc.) in the area which appears most active in terms of angiogenesis.

### 3.4.3. *Other applications*

Other diseases relating to modification of the vascularization of certain tissues can be diagnosed using perfusion sequences. It is also possible to monitor the impact of the treatment.

Thus, infectious or inflammatory diseases, of which one of the characteristics is hypervascularization, can be examined with perfusion MRI (abscesses, Crohn's disease, etc.).

# Chapter 4

# Functional MRI

## 4.1. Introduction

With MRI, it is possible to obtain maps of cerebral perfusion (see Chapter 3) by injection of gadolinium or continuous arterial spin labeling. However, to date, neither of these two techniques has been applied in the domain of neuroscience.

Another approach in MRI is fMRI (functional MRI), which – as we shall soon see – is based on local variations in blood flow and blood oxygenation: fMRI has become the main imaging method used in neurosciences.

In clinical – primarily neurosurgical – practice, this technique helps to locate the functional regions that are essential for motor- and linguistic skills, but also to assess the danger of post-operative deficiency.

## 4.2. Principle

### 4.2.1. *Origin of fMRI signal (BOLD effect)*

#### 4.2.1.1. *Metabolism of cerebral activity*

When the brain is performing a particular action, certain specific neurons are called upon, and transmit information (potential for actions) along their axons to their synapses. This triggers the release into the synaptic cleft of neurotransmitters which, in turn, by binding to the postsynaptic receptors, engender triggering or inhibiting potentials for action.

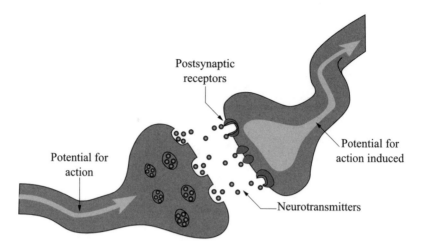

**Figure 4.1.** *Synaptic cleft and release of neurotransmitters*

These mechanisms require energy, delivered in the form of ATP, which needs glucose metabolized anaerobically by glycolysis and aerobically *using dioxygen*.

Glucose and dioxygen are carried by the arterial blood, and are extracted in the capillaries.

Thus, neuron activity causes an *increase in local dioxygen consumption*. This leads to an *immediate decrease in oxyhemoglobin concentration* (because of oxygen fixing by the hemoglobin) and an increase in deoxyhemoglobin (Hb).

### 4.2.1.2. *Neurovascular coupling*

One to two seconds after the start of neuron activity, the *cerebral blood flow* (CBF) *increases* very dramatically so as to serve the increased needs of glucose and oxygen of the active cortex: this response is known as neurovascular coupling.

### 4.2.1.3. *BOLD signal and hemodynamic curve*

In order to understand the apparition of a signal detectable with MRI, we need to take account of two important points:

– the magnetic properties of oxyhemoglobin and deoxyhemoglobin are different:

- oxyhemoglobin is created by the fixing of a molecule of dioxygen onto hemoglobin: this union is diamagnetic (as are all biological tissues) and therefore does not interfere with the external field (Bo) applied to these tissues,

- *deoxyhemoglobin* is paramagnetic, and *therefore interferes with the magnetic field applied*;

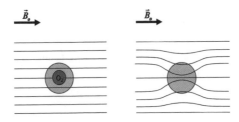

**Figure 4.2.** *Influence of oxy- and deoxyhemoglobin on magnetic field lines in close proximity*

– the neurovascular coupling detailed above is disproportionate to the requirements: the relative increase in blood flow (around 30%) is far greater than the increase in the oxygen consumption of the active neurons (around 4%);

– thus, we observe an increase in the concentration of oxyhemoglobin, with a resulting *drop in concentration of deoxyhemoglobin. (Here we consider the sum of the two concentrations to remain constant).*

Thus, the BOLD signal comprises three phases:

**1st phase**
As long as neurovascular coupling has not been activated, we note a rise in the deoxyhemoglobin concentration, and a consequent strengthening of the magnetic field ΔBo around the vessels, which then significantly interferes with the main field Bo.
This disturbance field shortens the transversal relaxation times of the protons in water molecules in the extravascular space: we see a decrease in the measured MRI signal.
This immediate response causes a rapid "dip" in the signal, which is a viable indicator of neuron activity, but it is inconstantly demonstrated.

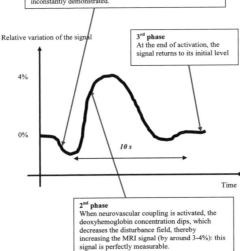

Relative variation of the signal

**3rd phase**
At the end of activation, the signal returns to its initial level

4%

0%

*10 s*

Time

**2nd phase**
When neurovascular coupling is activated, the deoxyhemoglobin concentration dips, which decreases the disturbance field, thereby increasing the MRI signal (by around 3-4%): this signal is perfectly measurable.

This effect of the variations in blood's magnetic properties with brain activation has been dubbed as "BOLD" (Blood Oxygen Level Dependent).

Generally, the initial dip in signal is not detected, and we obtain the following shape, which is called a *hemodynamic response*:

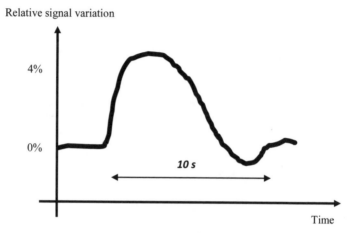

### 4.2.1.4. *Overview*

The diagram below offers a recap of the mechanism that leads to the signal growth when a hemodynamic response takes place.

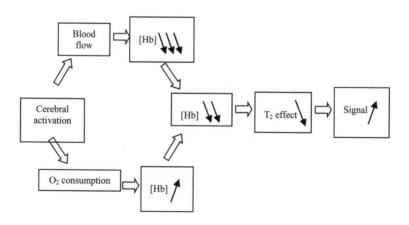

### 4.2.2. *Type of paradigm*

The technique of fMRI enables us to indirectly measure a subject's cerebral activity by detecting the BOLD signal. The goal of an fMRI examination is thus to show the areas of the brain which are activated when performing a specific task. In order to do this, the tester must:

– precisely define the question of interest (i.e. the cerebral areas under study);

– design the experimental paradigm, i.e. the sequence of stimuli applied over time, to find the sought areas.

#### 4.2.2.1. *Necessity of alternation between rest and activation*

The main problem lies in the fact that we cannot quantify the physiological parameters of activation (such as the oxyhemoglobin concentration or the blood flow) on the basis of the fMRI signal received. Therefore, it is impossible to determine an *absolute value* for the fMRI signal beyond which a voxel can categorically be said to belong to an active or inactive region.

The solution to this problem is to detect a *relative value* for the signal during a *state of activation in comparison to a state of rest,* which is called the *reference state*.

#### 4.2.2.2. *The two main types of paradigm*

Therefore, we need to alternate between situations of rest and activation.

The duration of the hemodynamic response is around 10 s (that is, the time taken for the signal to return completely to its initial value), so one possibility is to apply the events (the stimuli) approximately every 20 s.

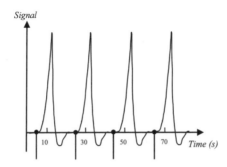

**Figure 4.3.** *BOLD signal due to a succession of stimuli (round-headed indicators)*

Drawbacks immediately become apparent:

– the subject spends a great deal of time doing nothing at all;

– the overall acquisition time is very long, although relatively few slices are actually acquired.

### 4.2.2.2.1. Block design paradigm

In order to deal with these issues, we can move the events closer together, sending them in packets of several stimuli. In this case, the hemodynamic responses created are amalgamated.

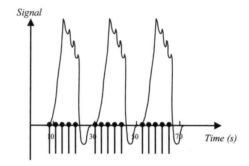

**Figure 4.4.** *BOLD signal when the stimuli are spaced a short time apart*

If the events are sufficiently close to one another (1-2 s), we obtain the shape of a block.

Each block is separated from the next by a rest period, whose duration is equal to that of one block (around 15-30 s).

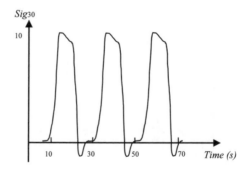

**Figure 4.5.** *Block-design BOLD signal*

These paradigms are the most commonly used. They offer many advantages:

– *Certain experiments are better suited for this type of paradigm.* Such is the case when we are looking at slow-changing cognitive processes, e.g. a change in attentional or emotional state.

– *The activated regions of the brain are clearly identified.* In order to determine the activated areas, it is necessary to perform statistical analysis (which we shall examine in section 4.3). The results of this analysis are clearer when the activation and rest periods are clearly identified, which is the case with the block design paradigm. Indeed, the rapid succession of stimuli in a single block causes a sustained response, and given that the task changes are relatively well spaced out over time, the rest period is sufficiently long for the signal to be able to return to its initial level.

– *Acquisition and processing are relatively easy.* With this type of paradigm, all of the tasks associated with an fMRI exam (construction of the paradigm, instruction of the subject and analysis of the results) become easier.

There are two main drawbacks to block design paradigms:

– inevitably, the subject being examined will become accustomed to the format of the test, and therefore anticipate the next event. These phenomena are superposed along with the cognitive tasks that the tester wishes to observe;

– this paradigm considers the average activity during the course of a block, and is therefore not suitable if we wish to study any variations in the subject's response during a block.

4.2.2.2.2. Event-related design paradigm

When the stimuli are quite separate from one another in time, we speak of an event-related design paradigm: in this case, the signal has the time to return to its equilibrium level between the application of two stimuli.

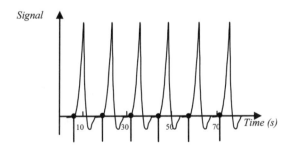

**Figure 4.6.** *BOLD signal recorded when using an event-related design paradigm*

Two advantages are immediately apparent:

– the phenomenon of anticipation mentioned previously in our discussion of the block-design paradigm no longer arises;

– the decline over time of each hemodynamic response can be precisely evaluated (because it can be evaluated on its own, rather than in combination with another response): it is therefore possible to extract and compare the characteristics of that response (rise time, peak intensity, etc.) depending on the stimuli presented.

The major drawback to this type of paradigm is that the small variation in signal caused by a single stimulus may yield relatively insignificant results in statistical tests (see section 4.3, thereby causing a loss of sensitivity in cerebral detection (see section 4.5.1.1.1).

One solution to help deal with this problem is to take an average of a number of responses (obviously all associated with identical stimuli).

4.2.2.2.3. Which paradigm should be used for which type of examination?

In a clinical context, we limit ourselves to obtaining whole functional zones (see section 4.6: Applications of fMRI), using a block-design paradigm.

In the context of neuropsychological study, it is preferable to use event-related design paradigms.

NOTE.–The way to choose an appropriate paradigm will be discussed again later on (see section 4.4.2.4.1), once we have analyzed the method of statistical analysis of the signal.

4.2.2.3. *How to choose the right reference state*

In the statistical study discussed in the next section, we are going to draw a comparison between the signal during the activation state and during the reference state. The reference period therefore needs to include all of the tasks occurring during the period of activation, with the exception of the particular task of interest (i.e. the task being studied).

This reference state may be more or less easy to define.

Thus, for instance, when we are looking for the areas controlling motor skills, the reference state will be one where the subject is making no voluntary movements.

On the other hand, when studying cognitive functions, the reference state will often be more difficult to define, given the multitude of processes that are involved.

### 4.2.3. *Type of sequence used in fMRI*

The sequences which are likely to exhibit the BOLD effect are those which are most sensitive to heterogeneity in the magnetic field: gradient echo (GE) sequences.

In "conventional" GE sequences, it takes a few seconds to obtain the image of one slice and several minutes to obtain an image of the whole volume of the brain. In view of the necessity of averaging several acquisitions of the same slice in order to obtain a sufficient SNR (see section 4.2.2.2.2), the use of such sequences would mean a prohibitively lengthy examination.

The technique we use, therefore, is echo-planar imaging (EPI). A slice can be imaged with a single excitation pulse, which considerably reduces acquisition time (the repetition time $T_R$ is around 50 ms with a 3T magnet with a spatial resolution of 64×64 for a slice; between 1.5 and 3 s for the whole of the brain).

#### 4.2.3.1. *Temporal resolution*

As can be seen from the shape of the hemodynamic response presented in section 4.2.1, the duration of the signal (start of rise, followed by a return to normal level) is around 10 s. This time-period therefore represents the temporal resolution of fMRI, because this is the length of time that we need to wait in order to observer any change in signal.

NOTE.–

– It is interesting to note here that the temporal resolution is limited not by the relaxation time of the spins, nor by the limitations of the machine, but rather by the very phenomenon which gives rise to the signal.

– This ~10 s delay of course does not mean that we cannot activate the voxels under study multiple times in order to take an average of the signal strength, thereby significantly enhancing the sensitivity in the cerebral detection performed in a statistical test.

#### 4.2.3.2. *Spatial resolution*

The spatial resolution is directly determined by the EPI sequence. In general, very fine spatial resolutions (<1 mm) can be obtained – particularly in mono-slice imaging. With multi-slice imaging, in order to avoid excessively long acquisition times, as a general rule we choose a resolution of around 3-4 mm, preferably identical in all three spatial directions.

### 4.2.3.3. *Example of a sequence*

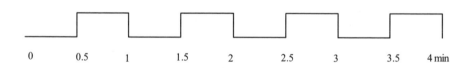

**Figure 4.7.** *Periods of activation (upper level) and inactivation (lower level) viewed with a conventional sequence*

Here, we have taken the example of a block-design paradigm with four periods of activation, intercalated with four rest periods. Each period lasts 30 s, during which time, 10 acquisitions of the same volume of brain space are taken.

## 4.3. Introduction to statistical analysis

### 4.3.1. *Why this study?*

#### 4.3.1.1. *Prior knowledge*

We have the signal from the particular voxel under examination, resulting from the implementation of a paradigm (alternating reference periods and activation periods).

NOTE.– For the moment, we shall not discuss the post-processing that needs to be applied to this signal (filters, etc.).

#### 4.3.1.2. *Aims*

Remember that fMRI examination is based on a possible difference between a state of rest and an activated state.

We can therefore distinguish two different objectives for an fMRI examination:

– Objective 1: certain studies have a relatively simple goal. For instance, we may wish to find out which voxels are more active during a particular task *than they are when at rest.*

– Objective 2: in more complex cases, we may also be interested in which voxels are more significantly activated by one particular task *than they are by a different one.*

In any case, following the execution of a paradigm (whose contents we shall discuss in greater details later on), we will have a signal for each voxel being examined. The objective is to use this signal to determine whether a given voxel belongs to the group of voxels sought in our initial question.

Thus, we are going to take the following two examples: we wish to know which cerebral regions are:

– activated by auditory perception of a stimulus (a scenario demonstrative of Objective 1);

– activated more strongly by processing words than numbers, when an auditory stimulus is received (demonstrative of Objective 2).

## ONE stimulus → determination of THE VARIOUS pixels sought

In order to deal with these problems, we propose to use the following paradigm: during each test, a subject is presented either with a word or a number; these stimuli are presented either visually (on a screen) or aurally (with headphones).

The instruction is to respond to the stimulus using a keyboard with two buttons (yes/no):

– if the stimulus presented is a word, answer "yes" if that word represents an animal;

– if the stimulus presented is a number, answer "yes" if that number is greater than 5.

Hence, this paradigm contains 4 conditions:

–"auditory word";

– "visual word";

– "auditory number";

– "visual number".

Thus, in the case of the second example, when we process the signal emitted by a given voxel during the application of the paradigm, we will want to know *"whether the condition 'auditory word' has a significant influence on the signal"*. More specifically, *"can the signal primarily be attributed to the condition 'auditory word'?"*

## ONE signal → determination of THE VARIOUS stimuli responsible for that signal

NOTE.– A voxel that is activated by the condition "auditory word" may also be activated by the condition "auditory number", for instance. This is why the word "primarily" is employed above.

### 4.3.2. *How should this analysis be done?*

In order to answer this question, we shall momentarily abandon the context of fMRI, to look at simpler examples which will help us to lay the foundations for our thinking process.

#### 4.3.2.1. *First sub-problem*

We know the production (in tons), number of hours worked and capital invested (in number of machines) for nine textiles companies.

We wish to find the answer to the following question: "Do the capital invested in a company and the amount of work done have a significant impact on that company's industrial production?"

The analogy between this problem and the study of an fMRI signal is as follows:

| fMRI | $\rightarrow$ | Initial example |
|---|---|---|
| Signal from the voxel | $\rightarrow$ | The company's production |
| Hearing of a word | $\rightarrow$ | The capital invested and/or work done |

Thus, we attempt to use a model to explain the variations in production on the basis of the other two variables (the capital invested and the number of hours worked).

The relation we are looking for can be expressed in the following form:

$$Y = f(X1, X2)$$

where Y represents production, X1 the capital invested and X2 the number of hours worked. Y is referred to as the explained variable or the response, whilst X1 and X2 are the explicative variables or factors.

#### 4.3.2.2. *Second sub-problem*

The above model is already relatively complex, because we are looking at the influence of two parameters on a single given variable.

Therefore, we shall choose to take one step back in terms of difficulty and examine the following problem. A tester wishes to calibrate a chromatograph. He wants to answer the following question:

"Does the ethanol content of a substance significantly influence the height of the peak measured by the chromatograph?"

### 4.3.3. *Simple linear regression technique*

In order to answer the above question, the test has performed ten tests using ten different levels of ethanol content (variable X) and obtained the following surfaces (variable Y).

| test | ethanol content (g/mL) | peak height (cm) |
|------|------------------------|------------------|
| 1 | 0.2 | 2.88 |
| 2 | 0.2 | 1.72 |
| 3 | 0.6 | 4.7 |
| 4 | 0.6 | 5.28 |
| 5 | 1 | 10.92 |
| 6 | 1 | 7.7 |
| 7 | 1.4 | 12.35 |
| 8 | 1.4 | 11.4 |
| 9 | 1.8 | 13.47 |
| 10 | 1.8 | 12.52 |

**Table 4.1.** *Set of values obtained in the ten tests*

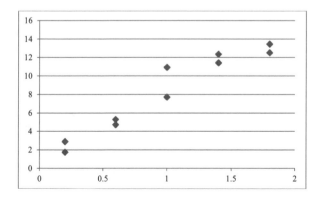

**Figure 4.8.** *Graph of peak height with varying ethanol content*

### 4.3.3.1. *Fundamental hypothesis*

In view of these results, the tester makes a hypothesis that the relation between temperature and yield is approximately linear, and wishes to obtain the parameters $\beta_0$ and $\beta_1$ for the corresponding straight line.

### 4.3.3.2. *Estimation of the parameters $\beta_0$ and $\beta_1$*

Given the sample presented in the above table, one might conceivably wonder when the straight line best represents the phenomenon; this means that we need to *estimate* the parameters $\beta_0$ and $\beta_1$ of the model by obtaining values $b_0$ (ordinate value at the origin of the line) and $b_1$ (slope) that are as close as possible to $\beta_0$ and $\beta_1$.

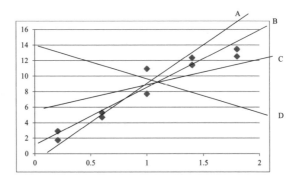

**Figure 4.9.** *Different possibilities for lines to best represent the points found experimentally*

It is clear that, of the various lines suggested, B is the one which is most appropriate, because it tends to minimize the differences between the observations and the graph. Thus, the estimators $b_0$ and $b_1$ of $\beta_0$ and $\beta_1$ will be found by minimizing the sum S of the squares of the (vertical) distances between the points and the plotted line:

$$S = \sum (Y - (b_0 + b_1 X))^2$$

We can determine $b_0$ and $b_1$ by differentiation S in relation to $b_0$ and $b_1$.

$$\frac{\partial S}{\partial b_0} = 0 \quad \text{and} \quad \frac{\partial S}{\partial b_1} = 0$$

We show that the solution is given by: $b_0 = \overline{Y} - b_1\overline{X}$ and

$$b_1 = \frac{\sum_{i=1}^{n}(X_i - \overline{X})(Y_i - \overline{Y})}{\sum_{i=1}^{n}(X_i - \overline{X})^2}$$

where the couples $(X_i, Y_i)$ are the couples of experimental points and

$$\overline{X} = \frac{1}{n}\sum_{i=1}^{n}X_i \quad \text{and} \quad \overline{Y} = \frac{1}{n}\sum_{i=1}^{n}Y_i$$

These formulae, when applied to our example, give us $b_0 = 1.22$ and $b_1 = 7.07$.

### 4.3.3.3. Quality of the estimation

Estimation of the parameters $\beta_0$ and $\beta_1$ enables us to calculate the estimate response for each observation. The difference between the observation made $Y_i$ and its estimation $\hat{Y}_i = b_0 + b_1 X_i$ gives the "residual" $\varepsilon_i = Y_i - \hat{Y}_i$.

Thereby, we can estimate the quality of the linear estimation in comparison to the observation made by summing together all of the residuals: $\sum_{i=1}^{n}\varepsilon_i^2$

This sum is averaged for the set of independent data (the number of acquisitions taken, less the two parameters from the model, which we call the number of degrees of freedom: see section 4.3.3.6).

Thus, we obtain the variance $\sigma^2 = \dfrac{\sum_{i=1}^{n}\varepsilon_i^2}{n-2}$

The quality of the model will be better when $\sigma^2$ is small.

NOTE.– The square root $\sigma$ of the variance is known as the standard deviation.

### 4.3.3.4. *Deviation between the estimated and exact values*

In the previous section, we performed ten successive acquisitions and were therefore able to estimate the parameters $\beta_0$ and $\beta_1$. If we now take ten more measurements, we shall obtain two new estimations, which will, *a priori*, be different.

This time, let us approach the problem differently: suppose we know that the substance's ethanol content is the determining factor for the height of the peak measured on the chromatograph.

Thus, we assume we have the relation $Y=\beta_0+\beta_1X+\varepsilon$

Thus, $\varepsilon$ could have any value whatsoever, but its statistical distribution ("the chances" of $\varepsilon$ having a particular value) must obey a normal law $N(0,1)$; its average value is zero and its standard deviation is 1.

*Mathematical point:* the standard deviation $\sigma$ is proportional to the width at half the height $\Delta\varepsilon$ of the curve: $\Delta\varepsilon=2\sqrt{2\ln2}\sigma$; thus, we can see that $\sigma$ characterizes the "spread" of $\varepsilon$).

We represent this distribution of $\varepsilon$ in the graph below:

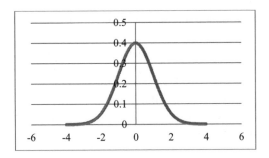

**Figure 4.10.** *Normal law*

Our aim is to evaluate the difference between the estimations $b_0$ and $b_1$ made in the wake of the acquisition of a certain number of experimental points and the "true" values of $\beta_0$ and $\beta_1$ .

With this goal in mind, we repeat a series of ten measurements, at the concentrations shown in the table above, 400 times. Thereby we obtain 400 estimations of the parameters $\beta_0$ and $\beta_1$ .

The 400 couples are represented below.

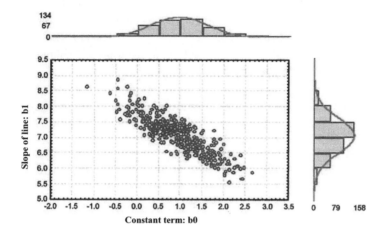

**Figure 4.11.** *Couples of estimators ($b_0, b_1$) obtained for each series of ten measurements*

In view of this graph, we can make the following observations:

– the points cloud obtained forms an ellipse;

– the center of that ellipse is the point ($b_{0ave} = 1$; $b_{1ave} = 7$).We can therefore say that on average (the mathematical term is "expectancy"), the estimators are free from bias, meaning that they are accurate: $E(b_0) = \beta_0$ and $E(b_1) = \beta_1$ ;

– the extent of the ellipse shows that each estimator has a certain variance, i.e. there is a certain dispersion of that estimator around its average value. Hence, if we were to project the couples ($b_0$; $b_1$) obtained onto each of the axes and plot a histogram for each of these parameters, we would obtain a symmetrical distribution that is *similar* to a *"normal"* law. More specifically, the higher the number of measurements taken (i.e. the larger the sample), the more closely the distribution will adhere to the normal law.

Thus, we can show that, if we hypothesize that the residuals are normal, $b_0$ and $b_1$ also follow a normal distribution, with the following variances:

$$V(b_0) = \sigma^2 \left( \frac{1}{n} + \frac{\overline{X}^2}{\sum_{i=1}^{n}(X_i - \overline{X})^2} \right) \quad \text{and} \quad V(b_1) = \sigma^2 \frac{1}{\sum_{i=1}^{n}(X_i - \overline{X})^2}$$

NOTE.– It is easy to understand the formulation of the variance of $b_1$: $b_1$ is of the "type" y/x. The possible errors relate to the estimation of y (not to x, which is perfectly well known). This accounts for the presence of the term $\sigma^2$ in $V(b_1)$.

(The error in estimation of $b_0$ and $b_1$ will be lesser when the points are aligned and therefore the sum of the residuals is small).

Thus, we can calculate the standard deviations $s(b_0) = \sqrt{V(b_0)}$ and $s(b_1) = \sqrt{V(b_1)}$.

We find $s(b_0) = 0.81$ and $s(b_1) = 0.7$.

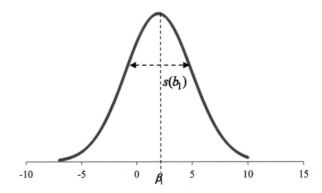

**Figure 4.12.** *Dispersion of the values of $b_1$ around the average $\beta_1$*

It must be understood that when taking a measurement (a sample), we have a certain probability of obtaining a particular value. That probability is proportional to the corresponding value on the curve.

### 4.3.3.5. *How are we to obtain a better estimation?*

Let us now turn our attention to the question of whether there is a way to influence the quality of these estimators, e.g. by attempting to obtain a smaller points cloud.

There are two factors which can influence the quality of estimators:

– the method of estimation;

– the quality of the sample.

*Estimation method*

The least squares method has one main strong point: it enables us to obtain non-biased estimators.

In addition, if the hypothesis that the residuals obey a normal law is vindicated, we can show that this method helps find the estimators with minimum variance out of all the possible estimators.

Thus, the least squares method is the best available estimation method.

*Sample quality*

There are a variety of factors which can be manipulated to affect the sample:

1. The less the variance in the residuals, the better the estimators will be. The choice of the parameter or parameters (in our case, ethanol content) to attempt to account for the evolution of the value under examination (the height of the peak) is therefore of crucial importance.

2. The variances of the parameters estimated are less when the size of the sample is larger ("the distribution is refined"). Thus, as might be expected, a *large sample* will yield better estimations.

If the number of samples is very high, we can assimilate the distribution of each of these two estimators to the distribution prescribed by the normal law.

Thus, the estimator $b_1$ obeys a normal law in terms of variance $V(b_1)$ and expectancy $\beta_1$.

### 4.3.3.6. *Confidence interval and Student's law*

In our study above, we sought to estimate the most accurate value of $\beta_1$ possible. In such a situation, we speak of a "confidence interval" when giving an interval which contains the exact value of $\beta_1$ with a certain degree of confidence.

How does one construct this confidence interval?

*Let us take the example of a variable x*, which obeys a normal law with *expectancy m and standard deviation $\sigma$.*

We then construct a new variable $(x-m)/\sigma$: this obeys a *reduced centered* normal law. We can show that as an ordinate, we obtain the probability density of each of the values of $(x-m)$.

Thus, the area defined for a given interval of the value (x-m), will give us the probability of obtaining a value in that interval.

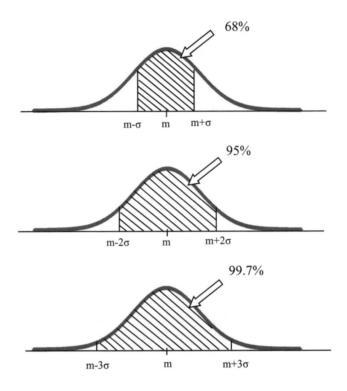

**Figure 4.13.** *Some examples of confidence intervals obtained, shown with their associated degree of confidence. (*CAUTION.– *the curves shown here are not those for the reduced centered normal law, but for a simple normal law)*

*Thus, let us attempt, having found an estimator $b_1$ and calculated the standard deviation $s(b_1)$, to give a confidence interval for $\beta_1$.*

The estimator $b_1$ does not obey a normal law, but is similar to it when the number of samples becomes large. In the same way as for the variable x, we construct a new variable: $\dfrac{b_1 - \beta_1}{s(b_1)}$ .

This variable conforms to Student's law (the "equivalent" of the reduced centered normal law obeyed by x).

The curve formed by Student's law gives the probability density of $\dfrac{b_1 - \beta_1}{s(b_1)}$.

Thus, for a given interval of $\dfrac{b_1 - \beta_1}{s(b_1)}$, which we shall notate as $\left[-t_\alpha ; t_\alpha\right]$, the area given by the curve of Student's law (the shaded area in Figure 4.15) will give the probability of obtaining a value for that interval after obtaining a sample.

The curve obtained is relatively similar to that obtained for the reduced centered normal law because $b_1$ obeys a law similar to the normal law.

More specifically (more accurately!), the curve obtained by Student's law is represented below:

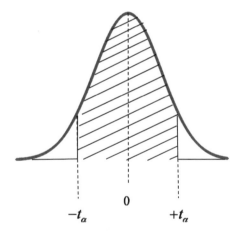

**Figure 4.14.** *Student's law*

We can give two equivalent interpretations of the shaded area of Student's curve. This area is proportional to:

– the probability of obtaining a value of $b_1$ within the interval $\left[\beta_1 - t_\alpha \cdot s(b_1); \beta_1 + t_\alpha \cdot s(b_1)\right]$;

— the probability of finding the sought value $\beta_1$ in the interval $\left[b_1 - t_\alpha \cdot s(b_1); b_1 + t_\alpha \cdot s(b_1)\right]$ after obtaining a given sample and calculating the value of $b_1$.

### 4.3.3.7....and in practice?

When we are faced with a given problem, the way of proceeding is the opposite:

— we define a threshold of acceptable risk $\alpha$ (in terms of percentage) of finding the value of $\beta_1$ *outside of* a particular interval (whose size is, for the moment, unknown);

— we define the size n of the sample (often determined by the acceptable duration of the experiment) which determines the number of degrees of freedom $\upsilon$ for the experiment (in our case: $\upsilon = $ n-2: see the comment below).

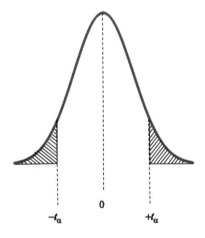

**Figure 4.15.** *The shaded area is proportional to the possibility of obtaining an estimation of $\beta_1$ outside of $\left[-t_\alpha; t_\alpha\right]$*

If we know $\alpha$ and $\upsilon$, we can deduce the value of $t_\alpha$, known as Student's coefficient; thus we know the breadth of the interval where there is a probability 1- $\alpha$ of finding $\beta_1$.

| | 0.50 | 0.20 | 0.10 | 0.05 | 0.02 | 0.01 | 0.005 | 0.002 | 0.001 | 0.0001 |
|---|---|---|---|---|---|---|---|---|---|---|
| 1 | 1.000 | 3.078 | 6.314 | 12.706 | 31.281 | 63.657 | 127.32 | 318.31 | 636.62 | 6366.2 |
| 2 | 0.816 | 1.886 | 2.920 | 4.303 | 6.965 | 9.925 | 14.089 | 22.327 | 34.599 | 99.992 |
| 3 | 0.765 | 1.638 | 2.353 | 3.182 | 4.541 | 5.841 | 7.453 | 10.215 | 12.924 | 28.000 |
| 4 | 0.741 | 1.533 | 2.132 | 2.776 | 3.747 | 4.604 | 5.598 | 7.173 | 8.610 | 15.544 |
| 5 | 0.727 | 1.476 | 2.015 | 2.571 | 3.365 | 4.032 | 4.773 | 5.893 | 6.869 | 11.178 |
| 6 | 0.718 | 1.440 | 1.943 | 2.447 | 3.143 | 3.707 | 4.317 | 5.208 | 5.959 | 9.082 |
| 7 | 0.71 | 1.41 | 1.895 | 2.365 | 2.998 | 3.499 | 4.029 | 4.785 | 5.408 | 7.885 |
| 8 | 0.706 | 1.397 | 1.860 | 2.306 | 2.896 | 3.355 | 3.833 | 4.501 | 5.041 | 7.120 |
| 9 | 0.703 | 1.383 | 1.833 | 2.262 | 2.821 | 3.250 | 3.690 | 4.297 | 4.781 | 6.594 |
| 10 | 0.700 | 1.372 | 1.812 | 2.228 | 2.764 | 3.169 | 3.581 | 4.144 | 4.587 | 6.211 |
| 20 | 0.687 | 1.325 | 1.725 | 2.086 | 2.528 | 2.845 | 3.153 | 3.552 | 3.850 | 4.837 |
| 30 | 0.683 | 1.310 | 1.697 | 2.042 | 2.457 | 2.750 | 3.030 | 3.385 | 3.646 | 4.482 |
| 40 | 0.681 | 1.303 | 1.684 | 2.021 | 2.423 | 2.704 | 2.971 | 3.307 | 3.551 | 4.321 |
| 50 | 0.679 | 1.299 | 1.676 | 2.009 | 2.403 | 2.678 | 2.937 | 3.261 | 3.496 | 4.228 |
| 60 | 0.679 | 1.296 | 1.671 | 2.000 | 2.390 | 2.660 | 2.915 | 3.232 | 3.460 | 4.169 |
| 70 | 0.678 | 1.294 | 1.667 | 1.994 | 2.381 | 2.648 | 2.899 | 3.211 | 3.435 | 4.127 |
| 80 | 0.678 | 1.292 | 1.664 | 1.990 | 2.374 | 2.639 | 2.887 | 3.195 | 3.416 | 4.096 |
| 90 | 0.677 | 1.291 | 1.662 | 1.987 | 2.368 | 2.632 | 2.878 | 3.183 | 3.402 | 4.072 |
| 100 | 0.677 | 1.290 | 1.660 | 1.984 | 2.364 | 2.626 | 2.871 | 3.174 | 3.390 | 4.053 |
| 200 | 0.676 | 1.286 | 1.653 | 1.972 | 2.345 | 2.601 | 2.839 | 3.131 | 3.340 | 3.970 |
| 300 | 0.675 | 1.284 | 1.650 | 1.968 | 2.339 | 2.592 | 2.828 | 3.118 | 3.323 | 3.944 |
| 500 | 0.675 | 1.283 | 1.648 | 1.965 | 2.334 | 2.586 | 2.820 | 3.107 | 3.310 | 3.922 |
| 1000 | 0.675 | 1.282 | 1.646 | 1.962 | 2.330 | 2.581 | 2.813 | 3.098 | 3.300 | 3.906 |
| infin. | 0.674 | 1.282 | 1.645 | 1.960 | 2.326 | 2.576 | 2.807 | 3.090 | 3.291 | 3.891 |

**Table 4.2.** *Values of $t_\alpha$ as a function of the number of degrees of freedom $v$ (vertically) and the risk threshold $\alpha$ (horizontally)*

COMMENTS.–

*– How many degrees of freedom are there?*

- One might reasonably ask this question, which justifies the relation $\upsilon = n - 2$ in the calculation of the number of degrees of freedom in the estimation of the values of $\beta_0$ and $\beta_1$.

- The n measurements taken when obtaining the sample are independent.

- The standard deviation of both the estimations compels us to posit that the averages (for the variables X and Y) in the population are equal to the averages in the sample. The upshot of this equality is that any two measurements out of the n available become dependent on the (n - 2) others, so the number of independent measurements at the moment of estimation is: $\upsilon = n - 2$. This number is the "number of degrees of freedom".

- This reasoning is absolutely universal; it can also be seen in the case of a multiple linear regression.

– Note that in the table shown above, the final row gives the figures for an infinitely large sample. In this case, $b_1$ obeys a perfect normal law, and the values obtained are those of a normal law table.

– When the sample size becomes too small, the uncertainty regarding the quality of the estimation becomes greater: the values of $t_\alpha$ become very large. As far as possible, we need to try to take a maximum number of measurements.

### 4.3.3.8. *Back to our example*

Let us return now to our original question:

"Does the ethanol content of a substance significantly influence the height of the peak measured by the chromatograph?"

If it is indeed shown to be the case that it does have a significant influence, the slope $\beta_1$ of the plot may have any value whatsoever, with the exception of absolute zero.

We then formulate the following *counter* hypothesis $H_0$: "$\beta_1 = 0$".

It is easier to determine whether a coefficient can be equal to a given value than whether it lies within a particular interval (here, $]-\infty;0[\cup]0;+\infty[$)

In order to find out whether the hypothesis $H_0$ is verified:

– we calculate the existing deviation between the estimation $b_1$ of $\beta_1$ and the tested value 0: $b_1 - 0$. We then compare that deviation to the estimated standard deviation $s(b_1)$ by calculating $t_{obs} = \dfrac{b_1 - 0}{s(b_1)}$ (the formulation $t_{obs}$ indicates that this relative deviation is observed or estimated on the basis of the sample obtained).

We obtain: $t_{obs} = \dfrac{7.07 - 0}{0.7} = 10.1$. This is equivalent to writing that $b_1 - 0 = 10.1 \cdot s(b_1)$.

– We give the confidence interval of $\beta_1$:

- For a risk threshold of $\alpha = 5\%$ (this value is usually chosen for fMRI purposes), and for n = 10 (which is so in our example, so $\upsilon = 10\text{-}2 = 8$), we have: $t_\alpha = 2.306$ (see Student's table above).

$\beta_1$ has a 95% chance of being in the interval $\left[b_1 - 2.306 \cdot s(b_1); b_1 + 2.306 \cdot s(b_1)\right]$.

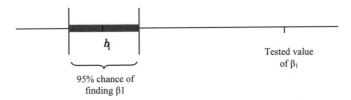

Tested value of $\beta_1$

95% chance of finding β1

We note that $\beta_1$ has only a 5% chance (at best) of being equal to the tested value (in this case 0).

The hypothesis $H_0$ can therefore be debunked, and its counter-hypothesis $H_1$: "$\beta_1 \neq 0$" can be accepted with a 5% risk threshold or a confidence level of 95%.

We now need to interpret our conclusion: the hypothesis that the ethanol content influences the peak height has only a 5% chance of being incorrect, whereas it has a 95% chance of being correct. This hypothesis will be either entirely correct or entirely wrong; however, we do not know which of these two scenarios we are dealing with!

We can only "make an educated guess", and say that "there is a 95% chance we are right in saying that the ethanol content influences the peak height".

### 4.3.3.9. Conclusion: general method of solution

From our investigation in the sections above, we can deduce the following general method, divided into three distinct phases: staging of the problem; estimation of the sought parameters; and conclusion.

| General technique | Application |
|---|---|
| *Staging of the problem* | |
| Formulate the question being asked | "Does the ethanol content of a substance significantly influence the height of the peak measured by the chromatograph?" |
| Identify the variables that play a part in the question: the explained variable (Y) and the explicative variable (X) | Y = peak height<br>X = ethanol content |
| *Using the linear model*, deduce an equation linking the variables now identified which offers a favorable solution to the question posed | $Y = b_0 + b_1 X$ |
| Formulate a hypothesis $H_0$ and its counter-hypothesis $H_1$ | $H_0$: "$\beta_1 = 0$"<br>$H_1$: "$\beta_1 \neq 0$" |
| Devise and implement an experimental protocol by which to acquire a sample | We obtain a set of experimental values ((Y; X) couples) |
| *Estimation of sought parameters* | |
| Estimate:<br>- the parameter at play in the hypothesis $H_0$<br>- the deviation of the model from the experimental reality: calculate the sum of the residuals | Estimation of $\beta_1$: thus, we obtain $b_1 = 7.07$ |

| | |
|---|---|
| - the variance of the model $\sigma^2$ <br> - the standard deviation of the sought parameter | Estimation of $s(\beta_1)$: thus, we obtain $s(b_1)=0.7$ |
| Set an acceptable risk threshold to validate $H_1$ (or to debunk $H_0$) <br> From this, deduce the deviation $t_\alpha$ | $\alpha = 5\%$ <br> $t_\alpha = 2.306$ |
| Estimate the relative deviation $t_{obs}$ | $t_{obs} = \dfrac{b_1-0}{s(b_1)} = 10.1$ |
| *Conclusion* | |
| Compare $t_{obs}$ against $t_\alpha$ <br> Conclude | $t_{obs} > t_\alpha$ : hypothesis $H_0$ invalidated (or $H_1$ validated) with a risk threshold of 5%: There is over a 95% chance that the ethanol content does influence the peak height. |

COMMENT.– In the case that $t_{obs} < t_\alpha$, we cannot reject $H_0$, but neither can we validate it.

### 4.3.4. *Multiple linear regression technique*

4.3.4.1. *Fundamental hypothesis and objectives*

Let us now return to our question about the influence of the capital invested in a company, and the hours worked, on its production.

We know the production (in tons), the number of hours worked and the capital invested (in number of machines) for nine textiles companies.

| Company | Work (hours) | Capital (machines/hours) | Production (100 tons) |
|---|---|---|---|
| 1 | 1100 | 300 | 60 |
| 2 | 1200 | 400 | 120 |
| 3 | 1430 | 420 | 190 |
| 4 | 1500 | 400 | 250 |
| 5 | 1520 | 510 | 300 |
| 6 | 1620 | 590 | 360 |
| 7 | 1800 | 600 | 380 |
| 8 | 1820 | 630 | 430 |
| 9 | 1800 | 610 | 440 |

We wish to answer the following question:

"Do the capital invested in a company and the amount of work done have a significant impact on that company's industrial production?"

We suppose the (possible) link between Y and the explicative variables to be linear: $Y = \beta_0 + \beta_1 X_1 + \beta_2 X_2$.

Y represents production, $X_1$ the capital invested and $X_2$ the number of hours worked.

There are now two variables $X_1$ and $X_2$which account for the values assumed by Y.

Hence, in order to answer the question posed above, we need to find out whether either $\beta_1 \neq 0$ or $\beta_2 \neq 0$.

As we saw in the previous section, the relation $Y = \beta_0 + \beta_1 X_1 + \beta_2 X_2$is unlikely to perfectly represent reality, for two reasons:

– as we saw with the previous example, experiments which are performed twice in conditions that we believe to be identical only very rarely yield the same result;

– a model is usually merely an approximation of a far more complex phenomenon. Here we propose to examine the influence of two variables $–X_1$ and $X_2$ –but it is entirely possible that we may have forgotten (or deliberately ignored) other parameters that are estimated to be less meaningful in accounting for the trend in industrial production Y.

Hence, as before, an error term ε will be introduced to explain the difference that exists between the model and the observation.

For each company, therefore, we can write the following relation:

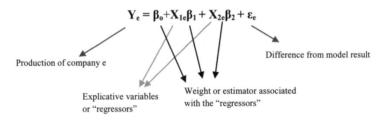

$$Y_e = \beta_0 + X_{1e}\beta_1 + X_{2e}\beta_2 + \varepsilon_e$$

Production of company e

Explicative variables or "regressors"

Weight or estimator associated with the "regressors"

Difference from model result

This is tantamount to writing the following matrix relation:

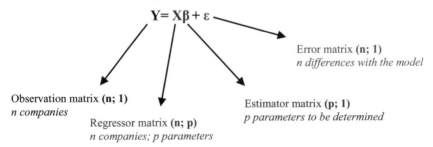

$$Y = X\beta + \varepsilon$$

Observation matrix **(n; 1)**
*n companies*

Regressor matrix **(n; p)**
*n companies; p parameters*

Estimator matrix **(p; 1)**
*p parameters to be determined*

Error matrix **(n; 1)**
*n differences with the model*

Initially, the aim is to determine **β** when we know **X**.

### 4.3.4.2. *Estimation of the matrix β*

NOTE.– $\beta_0$ represents a term which is constant to all the companies (similar to that which was introduced in the simple linear regression in the previous section).

The matrices corresponding to the data are:

$$
Y\begin{bmatrix} 60 \\ 120 \\ 190 \\ 250 \\ 300 \\ 360 \\ 380 \\ 430 \\ 440 \end{bmatrix}
\quad
X\begin{bmatrix} 1 & 1100 & 300 \\ 1 & 1200 & 400 \\ 1 & 1430 & 420 \\ 1 & 1500 & 400 \\ 1 & 1520 & 510 \\ 1 & 1620 & 590 \\ 1 & 1800 & 600 \\ 1 & 1820 & 630 \\ 1 & 1800 & 610 \end{bmatrix}
\quad
\varepsilon\begin{bmatrix} \varepsilon_1 \\ \varepsilon_2 \\ \varepsilon_3 \\ \varepsilon_4 \\ \varepsilon_5 \\ \varepsilon_6 \\ \varepsilon_7 \\ \varepsilon_8 \\ \varepsilon_9 \end{bmatrix}
\quad
\beta\begin{bmatrix} \beta_0 \\ \beta_1 \\ \beta_2 \end{bmatrix}
$$

The method used to obtain the estimators $\beta_i$ is the same as is used in the simple linear regression seen above; it is founded on minimizing the sum of the squares of the residuals $\varepsilon$.

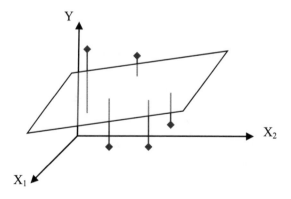

**Figure 4.16.** *Best planar representation of the experimental data*

The simple linear regression (see section 4.3.3) sought to draw a straight line "as closely as possible" through the points cloud representing the observations in the plane (X, Y). The same is true for multiple linear regression. However, the visual representation now becomes impossible, except in the example discussed here, where there are only two explicative variables $X_1$ and $X_2$: the objective then becomes to draw a plane "as closely as possible" through the points cloud representing the observations in the space $(X_1, X_2, Y)$.

We can make an analogy between the calculations for simple and multiple linear regression:

| Simple linear regression | Multiple linear regression |
|---|---|
| Minimization of $$S = \sum (Y - (b_0 + b_1 X))^2$$ | Minimization of $$S = \sum (Y - X\beta)^2$$ |
| Obtainment of the estimator: | Obtainment of the estimators: |
| 1. Calculation of $\overline{X}$ and $\overline{Y}$ <br><br> 2. Calculation of $$b_1 = \frac{\sum_{i=1}^{n}(X_i - \overline{X})(Y_i - \overline{Y})}{\sum_{i=1}^{n}(X_i - \overline{X})^2}$$ | 1. Calculation of the matrices $$(X^t X)^{-1} \text{ and } (X^t Y)$$ <br><br> 2. Calculation of $\hat{\beta} = (X^t X)^{-1} X^t Y$ <br><br> Matrix $(p; 1)$ |
| 3. Calculation of the variance of the residuals: $\sigma^2 = \dfrac{\sum_{i=1}^{n} \varepsilon_i^{\,2}}{n-2}$ | 3. Calculation of the variance of the residuals: $\sigma^2 = \dfrac{\sum_{i=1}^{n} \varepsilon_i^{\,2}}{n-p}$ |

| | |
|---|---|
| | where $\varepsilon_i = Y_i - (X\beta)_i$<br><br>$\sigma^2$ *is a scalar* |
| 4. Calculation of the variance and standard deviation of the estimators:<br><br>$V(b_1) = \sigma^2 \dfrac{1}{\displaystyle\sum_{i=1}^{n}(x_i - \overline{x})^2}$ | 4. Calculation of the variance and standard deviation of the estimators:<br><br>$V(\hat{\beta}_i) = \sigma^2 (XX')^{-1}_{i,i}$<br><br>and $s(\hat{\beta}_i) = \sqrt{V(\hat{\beta}_i)}$<br><br>$(XX')^{-1}_{i,i}$ *diagonal element of* $XX^t$<br><br>$XX^t$ *matrix(p; p)* |

*Mathematical point*

We seek to minimize $S = \sum(Y - X\beta)^2$.

We shall expand the matrix expression:

$$(Y - X\beta)^2 = (Y - X\beta)'(Y - X\beta) = Y'Y - \beta'X'Y - Y'X\beta + \beta'X'X\beta$$

Here we exploit the fact that $\beta'X'Y$ is a scalar, which implies that its transpose is equal to itself: $Y'X\beta = (Y'X\beta)^t = \beta'X'Y$.

Thus, $(Y - X\beta)^2 = Y'Y - 2\beta'X'Y + \beta'X'X\beta$.

Therefore, we need to minimize this expression in relation to $\beta$.

We calculate the matrix derivative of this expression in relation to $\beta$:

$$-2X'Y + 2X'X\beta$$

We posit this derivative as being equal to the null vector; the values obtained are those from the estimator of $\beta$ , which is notated as $\hat{\beta}$ :

$$X^t X \hat{\beta} = X^t Y$$

We can obtain $\hat{\beta}$ by multiplying both sides of the expression by $(X'X)^{-1}$:

$$\hat{\beta} = (X^t X)^{-1} X^t Y$$

### 4.3.4.3. *Back to our example*

Let us apply the method discussed above to our example:

1. Calculation of the matrices

$$(X^t X) = \begin{bmatrix} 9 & 13790 & 4460 \\ 13790 & 21672100 & 7066200 \\ 4460 & 7066200 & 2323600 \end{bmatrix}$$

$$(X^t X)^{-1} = \begin{bmatrix} 6.30477 & -0.007800 & 0.011620 \\ -0.007800 & 0.000015 & -0.000031 \\ 0.011620 & -0.000031 & 0.000072 \end{bmatrix}$$

$$(X^t Y) = \begin{bmatrix} 2530 \\ 4154500 \\ 1378500 \end{bmatrix}$$

2. Obtainment of the estimator

$$\hat{\beta} = \begin{bmatrix} \hat{\beta}_0 \\ \hat{\beta}_1 \\ \hat{\beta}_2 \end{bmatrix} = (X^t X)^{-1} X^t Y = \begin{bmatrix} -437.710 \\ 0.336 \\ 0.410 \end{bmatrix}$$

3. Calculation of the variance of the residuals:

Using this model, it is possible to give an estimation of the value of production for each company: $\hat{Y}_e = \hat{\beta}_0 X_{0e} + \hat{\beta}_1 X_{1e} + \hat{\beta}_2 X_{2e}$.

This enables us to find the variance of the residuals:

$$\sigma^2 = \frac{\sum_{i=1}^{n} \varepsilon_i^2}{n-p} = \frac{\sum_{i=1}^{n}\left(\hat{Y}_e - Y_e\right)^2}{9-3} = 532$$

4. Calculation of the variance and standard deviation of the estimators:

$$\sigma^2 \left(XX'\right)^{-1} = \begin{bmatrix} 3355.56 & -4.152 & 6.184 \\ -4.152 & 0.008 & -0.016 \\ 6.184 & -0.016 & 0.038 \end{bmatrix}$$

Thus:

$$s(\hat{\beta}_0) = \sqrt{V(\hat{\beta}_0)} = 57.93\,;$$

$$s(\hat{\beta}_1) = \sqrt{V(\hat{\beta}_1)} = 0.08966\,;$$

$$s(\hat{\beta}_2) = \sqrt{V(\hat{\beta}_2)} = 0.1961\,.$$

Let us now return to our original question:

"Do the capital invested in a company and the amount of work done have a significant impact on that company's industrial production?"

We can split this question into two distinct hypotheses:

| "Does the capital invested in a company have a significant impact on that company's industrial production?" | "Does the amount of work done in a company have a significant impact on that company's industrial production?" |
|---|---|
| Hypothesis $H_0$: "$\beta_1 = 0$" <br> Counter-hypothesis: $H_1$: "$\beta_1 \neq 0$" | Hypothesis $H_0$: "$\beta_2 = 0$" <br> Counter-hypothesis: $H_1$ : "$\beta_2 \neq 0$" |
| Estimation $\beta_1$: $b_1 = 0.336$ <br> Estimation of s($\beta_1$): $s(b_1) = 0.0896$ | Estimation $\beta_2$: $b_2 = 0.410$ <br> Estimation of s($\beta_2$): $s(b_2) = 0.1961$ |

| Risk threshold: | $\alpha = 5\%$; $t_\alpha = 2.306$ |
|---|---|
| Estimation of relative difference $t_{obs} = \dfrac{b_1 - 0}{s(b_1)} = 3.75$ | Estimation of relative difference $t_{obs} = \dfrac{b_2 - 0}{s(b_2)} = 2.09$ |
| $t_{obs} > t_\alpha$ <br><br> Hypothesis $H_0$ invalidated (or $H_1$ validated) with a risk threshold of 5%: <br><br> **There is a 95% chance that the capital invested in a company influences that company's industrial production.** | $t_{obs} < t_\alpha$ <br><br> We cannot debunk hypothesis $H_0$, but neither can we validate it. |

REMARK. –

We can see that:

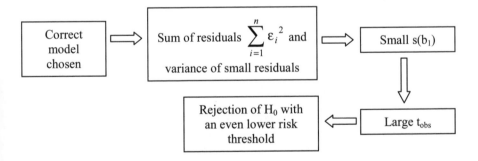

*A good model will enable us to validate a hypothesis $H_1$ with an even greater margin of confidence.*

## 4.4. Statistical analysis and fMRI

### 4.4.1. *Modeling of the fMRI signal*

#### 4.4.1.1. *Real signal and modeled signal*

In section 4.3.1.2, we put forward an experimental protocol that can be used to find the cerebral zones being sought. For each voxel, we obtain a signal Y (representing the hemodynamic response) containing n measurements (if we have acquired n scans): hence, this is a vector with size n.

CAUTION.– In reality, Y is a "manipulated" signal, i.e. one that has been smoothed, filtered, etc.

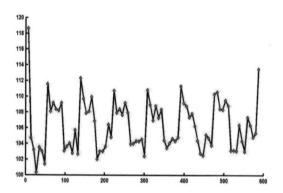

**Figure 4.17.** *Hemodynamic response plotted over time*

GOAL.– Our goal is to find out whether a particular voxel belongs to the sought area (in our example, the zone that is more strongly activated by the processing of words than of numbers, when an auditory stimulus is perceived).

Our task toward achieving this goal will entail reproducing the real signal Y as closely as possible by creating a simulated signal, constructed using theoretical responses to the different conditions of the protocol *(in our example: 4 conditions)*. Thereby, we can determine the contribution of each of these responses to the signal Y *(and thus discover whether the condition "auditory word" has had a predominant influence on the shape of that signal)*.

*3 stages of modeling:*

– Suppose that the protocol required c different conditions. We then make the hypothesis that the signal is linear (as we did for the examples not involving fMRI:

see section 4.3.3.1): if the response measured in the pixel being studied under the condition j is $Y_j$, then the real signal generated during the execution of the paradigm is:

$$Y = Y_1 + Y_2 + ... + Y_c$$

– Each response $Y_j$ will be replaced by its mathematical model, expressed in the form of a linear combination of theoretical time-dependent functions that are called *regressors* (or "basis functions") and which will be notated as $R_i$ (vectors of size n such as Y).*The form of these regressors will be discussed below.*

– In the previous two stages, therefore, we are seeking to obtain a simulation of the real signal in the form:

$$Y = \beta_0 R_0 + \beta_1 R_1 + ... + \beta_p R_p + \varepsilon$$

where:

- $\varepsilon$ (vector of size n) is the residual error and corresponds to the proportion of the data that is not accounted for by the model,

- $\beta_0 R_0$ is the average value of the signal: $R_0$ is a vector whose value is '1' for all measured points (also known as a "constant regressor").

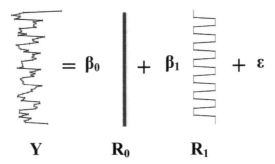

**Figure 4.18.** *Simulation with only two regressors, including the constant regressor*

Each parameter $\beta_i$ tells us the contribution of the regressor associated with the signal in a given voxel: the $\beta_i$ values which are significantly different rom 0 correspond to the regressors (and therefore to the conditions) which contribute greatly to accounting for the signal in the voxel.

A value β = 0 means that the signal in the voxel has not changed in relation to a situation where no stimuli were present. For this reason, it is necessary, as specified in section 4.2.2.1,to alternate between at least two different cognitive states during a single acquisition, so as to detect the areas of the brain where the signal has changed.

*Mathematical representation*: the set of p regressors is grouped into a matrix X called the "design matrix": X therefore has n rows and p columns.

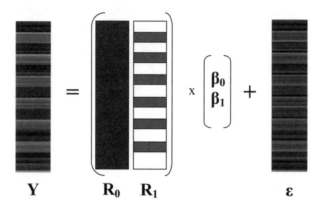

**Figure 4.19.** *Mathematical representation when there are only tworegressors, including the constant regressor*

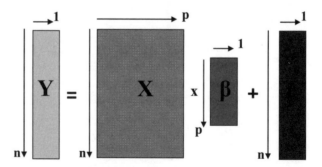

**Figure 4.20.** *Typical example*

*What we now need to know is, with regard to a particular sought functional zone:*

*– Which paradigm can be used?*

*– What is the shape of the matrix X associated therewith?*

*– Which types of regressors simulate the signal best?*

*Hereafter, we shall answer all these questions but not in the same order. We begin with the simplest of paradigms and therefore the simplest of matrices X, so as to illustrate the possible form of the regressors. Then, we shall go on to look at a more complex paradigm, which will require the construction of a matrix X that is also more complex.*

### 4.4.1.2. Example

4.4.1.2.1. An initial example to demonstrate the possible forms that a regressor may take

We construct a paradigm in which a single, visual stimulus is used.

We seek to find the regressor which most closely models the response.

*Type of regressor proposed:*

— the simplest form of the response caused is a "box-car response";

time

**Figure 4.21.** *Box-car response*

— this form is overly simplistic (the residuals it leaves when calculating the parameters $\beta$ are too great);

— as seen in section 4.2.1.3, the signal resulting from the BOLD effect is given by the hemodynamic response function (notated as "hrf");

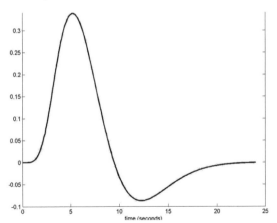

time (seconds)

**Figure 4.22.** *The hrf*

– if the condition is involved in the paradigm on multiple occasion, the appropriate regressor is given by the convolution of the stimulation signal by the hrf:

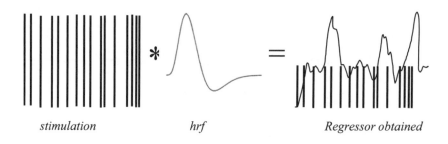

stimulation                     hrf                          Regressor obtained

**Figure 4.23.** *Convolution of the stimulation signal by the hrf*

*Conclusion:* we therefore model the signal with a single regressor hrf. The model is written as: $Y = \beta_0 R_0 + \beta_1 R_1$, where $R_0$ is the constant regressor and $R_1$ is the hrf.

*Our aim is to determine whether a voxel belongs to the visual area; in order to do so, we need to know whether the parameter $\beta$ corresponding to it is different to 0.*

4.4.1.2.2. Back to our initial question

Let us return to the problem posed in section 4.3.1.2. The paradigm is a specific succession of the four conditions: Auditory_Word, Visual_Word, Auditory_Number and Visual_Number.

For each condition, there is an associated parameter which shows the influence of that condition on the signal in the voxel under examination: thus, we have five parameters to evaluate for each pixel that we study: $\beta_1$, $\beta_2$, $\beta_3$ and $\beta_4$, and of course, $\beta_0$ (the parameter for the constant regressor).

*Construction of the matrix*

Software packages (such as SPM99) have a visualization of the matrix X, once the paradigm has been created.

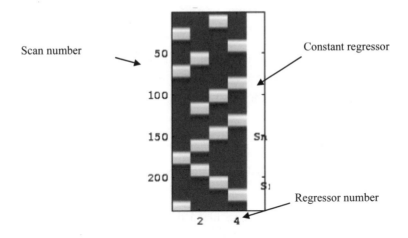

**Figure 4.24.** *Visualization of the matrix X with the regressors:*
*Auditory_Word, Visual_Word, Auditory_Number, Visual_Number*

Each white block comprises one or more hemodynamic response functions, separated by a given time interval (this interval is of course determined by the paradigm chosen).

The model is written thus:  $Y = \beta_0 R_0 + \beta_1 R_1 + \beta_2 R_2 + \beta_3 R_3 + \beta_4 R_4$, where the shape of the different regressors over time is given by each column of the matrix X.

We are interested in finding out which cerebral regions are more strongly activated by the processing of Words than by the processing of Numbers, only in case of auditory perception of the stimulus.

In order to determine whether a voxel is part of the regions sought, we need to know *whether $\beta_1 > \beta_3$.*

### 4.4.1.3. *Determination of the parameters and origin of the residual error*

In the first and second examples, we need to estimate the value of the parameters β, by solving the matrix equation which we have already examined in detail (see section 4.3).

With this is mind, we seek to minimize the residual error between the real signal and the modeled signal.

This method enables us to obtain:

– an estimation of the parameters β, in each voxel;

– the estimated signal in each voxel, or "reconstructed" signal based on the theoretical regressors and the estimated parameters;

– the residual error in each voxel – the difference between the data and the estimated signal.

*What, though, is the cause of the residual error?*

The signal measured by fMRI is not due only to the BOLD signal induced by the paradigm.

Thus, even if we think we have found the "perfect" model (although, is that really possible?), the real signal will never correspond perfectly to the modeled signal: this difference between the two signals is modeled by the residual error ε.

The origins of the residual error (besides poor modeling!) may be many, but are grouped into two main categories, depending on the way in which they can be modeled.

4.4.1.3.1. Low-frequency drifts

Low-frequency drifts include:

– artifacts from the patient's breathing;

– heartbeats;

– movements of the head.

These sources of error are modeled in a *deterministic* fashion by adding regressors into the matrix X (see section 4.4.1.4.1) or by applying a high-pass filter (see section 4.5.1.2).

Their "low-frequency" qualifier is obtained by opposition to the noise (see the next section), varying rapidly over time.

4.4.1.3.2. Noise

Noise is composed of:

– instabilities of the machine;

– thermal noise.

These random fluctuations are identified by rapid variations of the fMRI signal and are modeled probabilistically: for this reason, we perform a statistical study.

We model the noise in the form of "white noise": the noise at a given moment is independent of the noise at previous moments (we speak of "identically-distributed noise").

This means that the variance of the noise is the same regardless of the moment when it is measured. This latter point is extremely important, because all the variances made on the estimators are based on a single variance of ε (see section 4.3.3.4), which is time-independent.

### 4.4.1.3.3. Conclusion

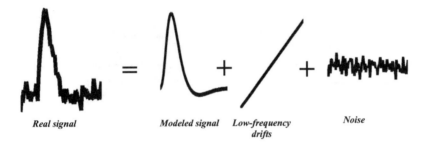

*Real signal*          *Modeled signal*   *Low-frequency drifts*          *Noise*

**Figure 4.25.** *Composition and origin of the fMRI signal (from [CIU 13])*

*In both the first and second examples given in section 4.3.1.2, the important thing is to determine these parameters with the maximum degree of precision (i.e. to keep ε as small as possible). Indeed (if we look at the second example), if the uncertainty is too great with regard to the parameters $\beta_1$ and $\beta_3$, we will be unable to determine whether one is larger than the other. In this case, we will be unable to respond with a minimum amount of certainty to the question being asked ("Which cerebral regions are more strongly activated by the processing of Words than by the processing of Numbers, only in case of auditory perception of the stimulus?"). Therefore, we need to construct the best possible model.*

### 4.4.1.4. How can we obtain the "best model"?

*The success criterion of a good model is to obtain the smallest possible residual error.*

For this purpose, we need to choose the right regressors – i.e. choose the regressors which are "truly" responsible, so that the model best represents the real signal. In the case of the example of industrial production seen above, one of the variables responsible for the result is the work that is done – not the weather outside the workshop!

In fMRI as well, we need to find the factors which truly are responsible for the signal observed.

4.4.1.4.1. Bring other regressors into play

The "low-frequency drifts" can be modeled by adding in so-called "regressors of no interest" (much like the constant regressor $R_0$). Generally, these regressors comprise a base of cosine functions with variable periods.

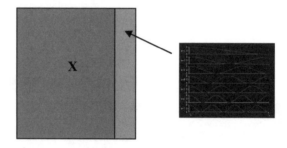

**Figure 4.26.** *Addition of regressors of the sinusoidal terms to the matrix...*

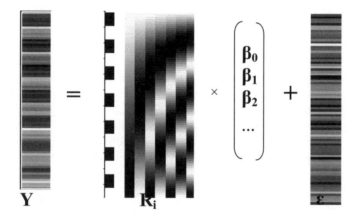

**Figure 4.27.** *...and visual representation of these terms in the general matrix relation*

Other functions may be used to best approximate the real signal (e.g. a gamma function, etc.).

It may happen that the errors stemming from noise are temporally correlated (therefore we can no longer make the hypothesis that we are dealing with white noise). The noise at a given time is related to the noise present at previous moments. This correlation does not cause an error in the estimation of the parameters β, but does adversely affect the estimation of the error made with regard to these parameters. Hence, it needs to be taken into account. Here, though, we shall not discuss the correction that consequently needs to be made (as the calculations are far more complex).

### 4.4.1.4.2. Adapt to the protocol

The hrf reaches its maximum point 5 s after the condition that has caused it. If the experimental condition requires a very variable length of reflection from the subjects (as is the case with a difficult task, for instance),modeling with only the hrf is insufficient. Voxels where the form of the signal is relatively different from the hrf alone (an earlier or later peak point, etc.) may be considered not to be activated although they are in fact activated (this is the phenomenon of "false negatives").In such a case, we can continue to use the hrf but accompanied by its derivative. Thus, we model the condition using two regressors instead of one (and therefore obtain two parameters).

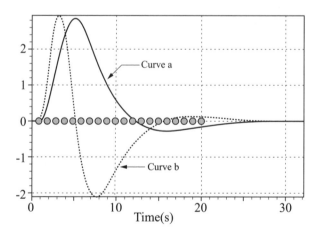

**Figure 4.28.** *The HRT function (curve a) and its derivative (curve b)*

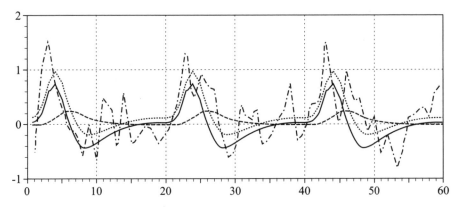

**Figure 4.29.** *Contribution to the simulation of the signal (curve a) made by the hrf (curve b), its derivative (curve c) and their sum (curve d)*

When a condition is repeated multiple times in a row, we consider that, in each test, that condition makes exactly the same contribution to the signal (in the simplest case, an hrf).

The contributions made by these successive tests are added together. If these contributions are sufficiently similar to one another (as happens with the block-design paradigm: see section 4.2.2.2.1), the resulting signal takes the form of a "box-car" function, smoothed and temporally shifted in relation to the first moment when the condition presents itself.

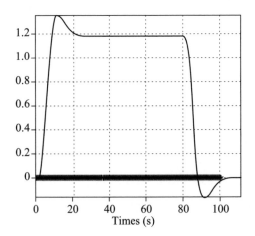

**Figure 4.30.** *"Box-car" function*

### 4.4.1.4.3. Conclusion

The choice of regressors is therefore of crucial importance in obtaining the "best possible" model of the real signal.

In general, the following guidelines are valid:

– if the chosen paradigm involves cerebral zones of whose location we are reasonably certain (e.g. the auditory or visual areas, etc.), then we need to limit ourselves to a few, simple regressors;

– if the paradigm is complex, and we have only very little information about the location of the cerebral zones which will be activated, we need to introduce more flexibility into the model, and thus offer a wide range of regressors, or indeed use more complex regressors;

– if all we have at our disposal are assumptions about the zones that will be activated, we can then perform two different simulations using the same paradigm:

- a simple simulation (few regressors and hrf alone),

- a more complex simulation (a large number of regressors, derivative functions, etc.).

We then compare the regions shown as activated by each of these two models: the more complex model will only be retained if the number of activated regions it contains is larger than the simpler model.

CAUTION.– Simple models are easier to interpret from a cognitive point of view. For that reason, they should be favored as much as possible!

NOTE.– In order to decrease the $\varepsilon$, we *also* need to choose an appropriate protocol: we shall discuss this later on (see section 4.4.2.4).

*By obtaining the parameters $\beta$ along with their associated uncertainty, we should be able to answer the questions posed with a degree of certainty defined by a given percentage of confidence: this is the objective of the statistical test.*

### 4.4.2. *Statistical test*

Here, we are going to be able to determine the voxels belonging to the target areas, in the two example problems posed above, with a given percentage of confidence.

*4.4.2.1. The t test*

4.4.2.1.1.  Resolution of the first example

We need to formulate a hypothesis $H_0$ and its counter-hypothesis $H_1$, such as those proposed in the two examples in section 4.3

$$H_0: \text{``}\beta_1 = 0\text{''} \quad \text{and} \quad H_1: \text{``}\beta_1 \text{ different to } 0\text{''}$$

$H_0$ is tantamount to writing $\displaystyle\sum_{i=1}^{c+1} c_i\beta_i = 0$, or in matrix form, $C'\beta = 0$

where $C' = \begin{bmatrix} 1 \\ 0 \\ 0 \\ 0 \end{bmatrix}$.

With the problem formulated thus, we can see that we need to carry out a test on a linear combination of parameters: in fMRI terminology, this is called "a contrast", which is written as $C = \begin{bmatrix} 1 & 0 & 0 & 0 \end{bmatrix}$.

The hypothesis $H_0$ which we wish to test leads us to perform a Student test: the hypothesis $H_0$ will be rejected, and therefore the voxel will be considered activated with a given percentage of confidence, which establishes the value of the parameter t.

The whole of the methodological process relating to this test was examined in section 4.3: choice of a percentage of confidence, determination of the parameter in Student's table, and conclusion.

As stated in the concluding remark in section 4.3.4.3, we can see that the percentage of confidence with which we can reject $H_0$, and therefore accept $H_1$, will be greater when $\varepsilon$ is small.

4.4.2.1.2. Resolution of the second example

Here, $H_0: \text{``}\beta_1 = \beta_3\text{''}$    and    $H_1: \text{``}\beta_1 > \beta_3\text{''}$

$H_0$ can also be written in matrix form: $C^t\beta=0$ with $C^t = \begin{bmatrix} 1 \\ 0 \\ -1 \\ 0 \end{bmatrix}$

Once again, we need to perform a Student test to invalidate or vindicate the hypothesis $H_0$.

### 4.4.2.2. Are there other tests?

#### 4.4.2.2.1. Debunking of the t test

If other types of tests are needed, the reason is that the t test is unable to validate or disprove all the hypotheses made by the experimenter. We shall now give a few examples of this.

Suppose we wish to perform a search identical to that outlined in the second example. This time, though, the model chosen is a little more complicated, because in addition to the hrf regressors $\beta_1$, $\beta_2$, $\beta_3$ and $\beta_4$, we introduce the derived hrf's (improvement of the model): $\beta_1'$, $\beta_2'$, $\beta_3'$ and $\beta_4'$. Each condition is therefore no longer modeled by a single regressor, but rather by two. (For instance, the condition "Auditory_Word" is weighted by two regressors $\beta_1$ and $\beta_1'$).

The expression of the problem posed in the form of a hypothesis is:

$H_0$: "$\beta_1=\beta_3$ **and** $\beta_1'=\beta_3'$" and $H_1$: "$\beta_1>\beta_3$ **or** $\beta_1'>\beta_3'$"

We can no longer resolve this question with a t test, because the test relates to two different conditions.

Now suppose that when putting a model in place (i.e. when choosing regressors which are intended to recreate the real signal as faithfully as possible), we wish to test the relevance of certain potential regressors.

Thus, in the above case, we may question the advantage of introducing the parameters $\beta_1'$, $\beta_2'$, $\beta_3'$ and $\beta_4'$.

The expression of the problem posed in the form of a hypothesis is:

$H_0$: "$\beta_1'=0$ **and** $\beta_2'=0$ **and** $\beta_3'=0$ **and** $\beta_4'=0$" and $H_1$: "$\beta_1'\neq0$ **or** $\beta_2'\neq0$ **or** $\beta_3'\neq0$ **or** $\beta_4'\neq0$"

Here again, the test relates to a number of conditions, so we cannot resolve the question using a t test.

Another example lies in the addition of regressors of no interest (to help model the low-frequency drifts: see section 4.4.1.4.1. Is it necessary to introduce these new regressors in order to improve the model?

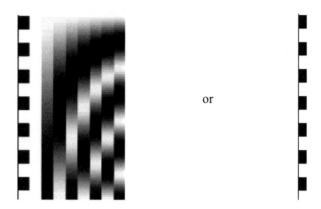

or

**Figure 4.31.** *Representation of the matrix X with (left) and without (right) the regressors of no interest*

*Conclusion:* for these particular problems, but for other ones as well, we need to employ a different test: an F test. We shall not go into detail about the F test as we did for the t test (to do so would be pointless, as the reasoning is largely identical).

4.4.2.2.2. Should we use a t test or an F test?

We use t contrasts to test, in each voxel of the brain, whether the activation associated with a given condition is greater than that associated with another condition.

F contrasts are used to test multiple regressors at once (conditions that involve multiple regressors or improvement and modification of the model).

4.4.2.3. *Is it always possible to find a specific functional zone?*

To determine a given function zone is to estimate a contrast (see section 4.4.2.1).

However, a contrast is based on the estimation of parameters.

Thus, asking the question about the feasibility of a cognitive study (looking for a specific functional zone) is tantamount to asking whether or not it is possible to evaluate the parameters $\beta$ unambiguously.

4.4.2.3.1. Step one: estimation of the parameters β

*An initial example*: in order to illustrate the difficulty of estimating the parameters β with precision, we are going to consider the case of a model (which will probably never be tested, given the lack of interest it holds) with two identical regressors: $R_1 = R_2$, whose corresponding parameters are $\beta_1$ and $\beta_2$.

The modeled signal therefore is $Y = \beta_1 R_1 + \beta_2 R_2 = (\beta_1 + \beta_2) R_1$: as the same condition is being tested twice, any solution of the type $\beta_1 + \beta_2 =$ constant will be valid.

There are an infinite number of possible couples $(\beta_1; \beta_2)$: the parameters $\beta_1$ and $\beta_2$ are not estimable uniquely.

*A "more commonplace" example:* while the scenario discussed above is relatively extreme and therefore very rare, the same cannot be said of a great many matrices X constructed using "conventional" paradigms.

Consider the following matrix of regressors: $X = \begin{bmatrix} 1 & 0 & 1 \\ 0 & 1 & 1 \\ 1 & 0 & 1 \\ 1 & 1 & 1 \end{bmatrix}$

We can easily determine a relation between these regressors: $R_3 = R_1 + R_2$. *Hence, these regressors are not independent.*

Suppose the parameters $\{\beta_1; \beta_2; \beta_3\}$ are obtained by linear regression to model the real signal. The modeled signal is therefore: $Y = \beta_1 R_1 + \beta_2 R_2 + \beta_3 R_3$

However, the modeled signal can also be written in the following form (if a is a constant):

$Y = (\beta_1 + a)R_1 + (\beta_2 + a)R_2 + (\beta_3 - a)R_3$

Thus, the regressors $\{\beta_1 + a; \beta_2 + a; \beta_3 - a\}$ are just as adequate as parameters for the model. That is to say that the parameters cannot be estimated in only one way.

*Conclusion*: the parameters β are not all estimable in a single way if the regressors in the matrix X are not linearly independent (if a given regressor is a combination of the others). Software packages for fMRI show any correlation between the regressors taken two by two, using white boxes (non-correlated regressors or "orthogonal" regressors), black boxes (completely correlated regressors) or gray boxes ("intermediary" correlation).

The more the regressors overlap, the greater is their correlation.

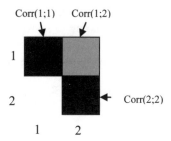

**Figure 4.32.** *Different correlations between regressors*

4.4.2.3.2. Step two: estimation of the contrasts

*i) Two examples*

Looking again at the scenario we have just examined, we can see that:

– the contrast $C = \begin{bmatrix} 1 & 0 & 0 \end{bmatrix}$ (corresponding to the hypothesis $H_0$: "$\beta_1=0$"). The value of $\beta_1$ cannot be obtained (or estimated), because $\beta_1$ can only be found to a particular degree of uncertainty – i.e. to a certain constant);

– the contrast $C = \begin{bmatrix} 1 & -1 & 0 \end{bmatrix}$ (corresponding to the hypothesis $H_0$: "$\beta_1=\beta_2$") is estimable (the constant a is identical for both these parameters, and is therefore eliminated).

NOTE.– As we can see here, this contrast is estimable but the parameters $\beta_1$ and $\beta_2$ are not!

*ii) Generalization*

A contrast is estimable if:

– the sum of the coefficients involved in that contrast is zero;

– the contrast which contains a single "1", associated with the estimable parameter $\beta$, is itself estimable.

*iii) A "less mathematical" approach*

We can understand this inability to estimate parameters or contrasts in a more common-sense light: the paradigms in fMRI are based on the *difference* in the subject's response between a period of rest and a period of activity. If the regressors used to model the signal "overlap", which is to say that they are correlated, we may attribute part of the activation signal during the period of activity to regressors describing the rest period: in other words, we may "miss" some of the signal and fail to detect certain activated pixels.

4.4.2.4. *A recipe for a successful functional study*

4.4.2.4.1. Before acquisition: how are we to construct an "appropriate paradigm"?

In the foregoing sections, we have already touched on (see section 4.2.2.2) the advantages and disadvantages of the two main types of possible paradigms (block design or event-based design). Now that we are familiar with the method for statistical treatment of the signal, we can gain a fuller understanding of the influence of the paradigm, on the responses given but also on the reliability of those responses.

In the construction of the paradigm, a few important points need to be borne in mind:

– clearly defining the types of functional zones sought: this will help us decide which comparisons to make between the parameters, and therefore which conditions are associated with those parameters;

– choosing the type of paradigm on the basis of the requirements of the study,

    - blocks: means we do not have to switch too frequently between the tasks,

    - single event: helps avoid monotony in the execution of the tasks,

– orthogonalizing the matrix of regressors as fully as possible: mixing the tests corresponding to the different conditions in a random order so that the regressors are not correlated.

4.4.2.4.2. After acquisition: how do we construct the matrix of regressors?

Here we wish to show that, with a paradigm set and the acquisition performed, an appropriate construction of the matrix of regressors helps provide an answer to many cognitive questions.

We propose the following paradigm: in each test, the subject is presented with a word, belonging to one of three possible categories ($M_1$, $M_2$, $M_3$); these words are presented either visually (on a screen) or aurally (through headphones).

Thus, we have six different conditions, represented summarily by the diagram below:

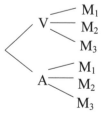

**Figure 4.33.** *Organigram of the stimuli present in the paradigm*

A specific response is required, depending on the stimulus that is presented.

4.4.2.4.3. An initial construction of the matrix of regressors

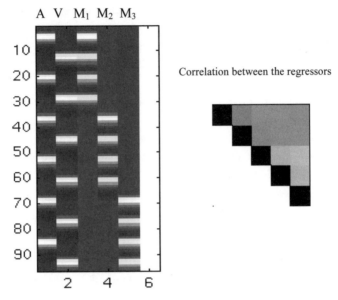

**Figure 4.34.** *One possible configuration for the matrix X*

Using this form of the matrix of regressors, we know:

– The regions where the activations associated with different categories of word are not the same (irrespective of the form of the stimulus – visual or auditory):

- The regions sought might, for instance, be those where $M_1 > M_2$.

- The corresponding contrast is $C_1 = \begin{bmatrix} 0 & 0 & 1 & -1 & 0 & 0 \end{bmatrix}$ (The final "0" corresponds to the constant regressor).

– The regions where the activations associated with different ways of presentation of the stimuli (visual or auditory) are not the same (irrespective of the type of word: $M_1$, $M_2$ or $M_3$):

- The regions sought might, for instance, be those where $A > V$.

- The corresponding contrast is $C_2 = \begin{bmatrix} 1 & -1 & 0 & 0 & 0 & 0 \end{bmatrix}$.

However, this matrix of regressors presents two major drawbacks:

– the orthogonality of the regressors is far from perfect;

– it is impossible to find an answer to some cognitive questions. Thus, the link between the form of the stimulus (auditory or visual) and the category of words cannot be examined.

### 4.4.2.4.4. A possible improvement

We can transform the above matrix of regressors into a new matrix:

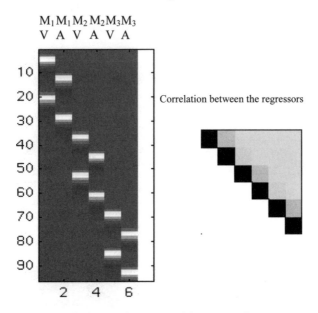

**Figure 4.35.** *New configuration of the matrix of regressors*

As in the case of the previous matrix, it is possible to test:

– the influence of the form of the stimuli (auditory or visual).The corresponding contrast is $C_3 = \begin{bmatrix} 1 & -1 & 1 & -1 & 1 & -1 & 0 \end{bmatrix}$;

– the influence of the form of the category of word. The corresponding contrast is $C_4 = \begin{bmatrix} 1 & 1 & -1 & -1 & 0 & 0 & 0 \end{bmatrix}$.

The improvements made here in comparison to the first matrix are:

– the possibility of testing the connection between the form of the stimulus (auditory or visual) and the category of words. The corresponding contrasts are $C_5 = \begin{bmatrix} 1 & -1 & -1 & 1 & 0 & 0 & 0 \end{bmatrix}$, $C_6 = \begin{bmatrix} 0 & 0 & 1 & -1 & -1 & 1 & 0 \end{bmatrix}$ and $C_7 = \begin{bmatrix} 1 & -1 & 0 & 0 & -1 & 1 & 0 \end{bmatrix}$:

- these contrasts enable us to look for the functional zones where the activation with a specific category of word is dependent on the form of the stimulus used.(For instance, $C_5$ looks for functional zones where the words $M_1$ are activated by a visual stimulus and the words $M_2$ are activated by an audio stimulus),

– the fact that the regressors are practically all orthogonal.

### 4.4.2.5. Activation map and choice of a statistical threshold

### 4.4.2.5.1. Creation of an activation map

The ultimate objective of our study is to be able to see the cerebral zones that are activated for a given cognitive issue. The steps in statistical analysis are as follows:

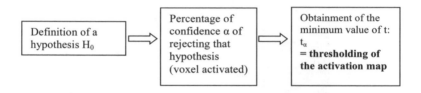

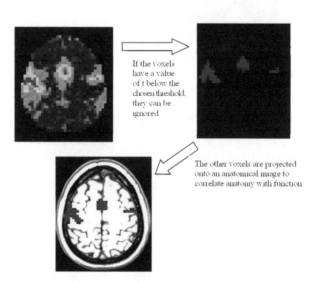

**Figure 4.36.** *Creation of an fMRI image by thresholding. For a color version of the figure, see www.iste.co.uk/perrin/MRITech.zip*

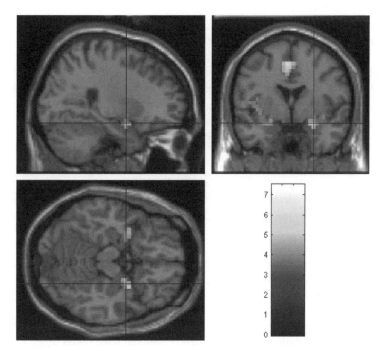

**Figure 4.37.** *Cerebral areas activated by the perception of smells. For a color version of the figure, see www.iste.co.uk/perrin/MRITech.zip*

### 4.4.2.5.2. How are we to choose the statistical threshold?

The choice of statistical threshold depends on the cognitive areas that we are trying to find.

We can distinguish between two main categories:

– *primary areas* (particularly the motor and visual centers) which can be detected with a high statistical threshold for t (corresponding to a risk threshold of less than 0.1%);

– *associative areas* (the secondary or tertiary areas which process the information received from several primary areas), where the statistical threshold is lower (corresponding to a risk threshold of around 1%).

### 4.4.2.6. *The problem with multiple tests*

### 4.4.2.6.1. Passage from one test to multiple tests

Hitherto, we have focused solely on the possibility of activation or non-activation of a single voxel with a given paradigm.

In order to create an activation map, we need to perform a large number of tests, which leads to a problem in "group analysis", as we shall soon see.

PROBLEM.– We take the acquisition of a cerebral volume of $\mathbf{q}$ = 10,000 voxels. Each voxel represents one statistical value of t which we wish to determine (obtention of $t_{obs}$).

In our case, we shall take the number of degrees of freedom to be $\upsilon$ = 40.

REMINDER.– (see section 4.3.3.6):

*This number is determined on the basis of the number $\mathbf{n}$ of temporal measurements carried out for each voxel, and the number $\mathbf{p}$ of regressors chosen to construct the model.*

*The number of degrees of freedom is then given by $\mathbf{n\text{-}p}$.*

For a voxel:

– The risk threshold, or the probability of obtaining a false positive for the particular voxel in question, is set at $\alpha$=10%. Using the table given in section 4.3.3.7 (Figure 4.17.), we obtain the value $t_\alpha$ = 1.684.Thus, we can state that if $t_{obs}$ >1.684, there is a 10% chance that the voxel considered activated will actually be a false positive.

– Let us now increase the number of tests performed:

   - For 1 sample, the probability of obtaining 1 positive voxel is 10%;

   - For 5 samples, the probability of obtaining 1 positive voxel is 50%

   - For 10 samples, the probability of obtaining 1 positive voxel is 100%.

Hence, inevitably, there will be at least one false positive voxel out of the ten voxels tested.

*The probability of obtaining a given value during the analysis is multiplied by the number of voxels on which the model is tested. The consequences are significant: with the 10,000 tests being performed, statistical analysis will, on average, cause 1,000 false positive voxels!*

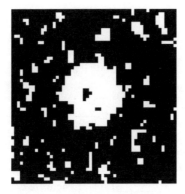

**Figure 4.38.** *Risk threshold: 10%.There are 10% false positives in this image.*
*(From [HIL 08])*

4.4.2.6.2. Bonferroni correction

The probability needs to be "corrected" – i.e. it needs to take account of the number of tests performed: here the number of voxels analyzed.

*The so-called Bonferroni procedure is the simplest way of performing this correction by dividing the chosen statistical threshold by the number of tests carried out.*

*Look at the impact of the Bonferroni correction with the following example.*

We wish to work with a 10% risk of error:

– For 1 sample with 40 degrees of freedom:

  - the probability of obtaining 1 false positive voxel is 10%;

  - the corresponding statistic t is $t_\alpha = 1.684$.

– For a volume of brain matter q = 10,000 samples, with 40 degrees of freedom:

  - the new statistical threshold (or "corrected probability") is:
  10%/10,000 = 0.0010% = 0.000010 for each voxel!

  -    the    probability    of    obtaining    1    false    positive    voxel
is:0.0010% × 10,000 = 10%;

  - the corresponding statistic t is $t_\alpha = 4.38$;

  - this threshold is clearly far more restrictive than the previous one.

Another way of looking at this is to say that in 100 images containing 10,000 pixels, there will be 10 images that contain a false positive pixel.

Thus, the formulation of the risk changes when the Bonferroni correction is applied:

| |
|---|
| ***BEFORE*** |
| Statistical threshold: α |
| **There is an α% chance that the** |
| **voxel will be false positive** |

| |
|---|
| ***AFTER*** |
| New statistical threshold: α/n |
| There is an α/n% chance that the voxel |
| will be false positive |
| There are α% of images containing |
| **one false positive voxel** |

4.4.2.6.3. Dependency on the tests and "new" Bonferroni correction

Let us apply the Bonferroni correction to our case study and observe the activated pixels.

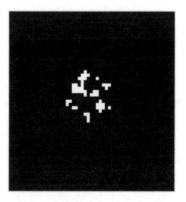

**Figure 4.39.** *Bonferroni correction and activated pixels. (From [HIL 08])*

The number of voxels shown as being activated is very low. Thus, while the Bonferroni correction is valid in all cases, *it seems excessively severe in certain cases*.

In order to understand what these particular cases are, we are going to take a simulated image of a slice (same number of voxels: 10,000).

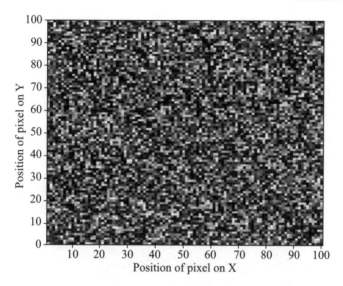

**Figure 4.40.** *The values of the voxels are obtained using random numbers taken from a normal law (see Figure 4.43). The white pixels are positive with the greatest probability. (From [BRE 03])*

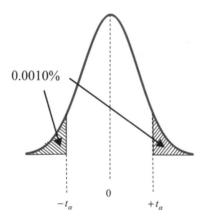

**Figure 4.41.** *Normal law. The shaded area is proportional to the probability of obtaining a value for $t_{obs}$ greater than $t_\alpha = 4.38$*

Remember that Student's law approximates a normal law when the number of degrees of freedom tends toward infinity. Thus, these random numbers can be matched to a random distribution of values of $t_{obs}$ on the Student curve.

With the Bonferroni correction shown previously (taking the statistical threshold t as $t_\alpha = 4.38$), we expect 10 out of 100 images to contain one or more pixels whose $t_{obs}$ value is greater than 4.38.

Using the previous image, let us carry out the following operations:

– divide the image into $10 \times 10$-voxel squares;

– in each square, replace the value of each voxel with the average of the 100 values in the whole of the square.

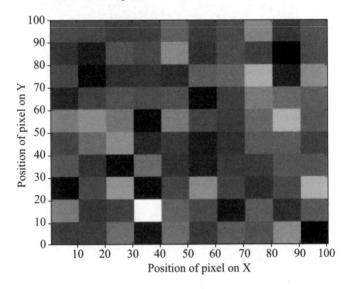

**Figure 4.42.** *Grouping of pixels into 10×10 squares to introduce a correlation between them. (From [BRE 03])*

There are still 10,000 voxels in this image, but now there are only $10 \times 10 = 100$ independent values. Hence, in reality, there are now only 100 tests that need to be performed.

Thus, if we keep the previous Bonferroni correction (0.0010%), the probability of obtaining one false positive voxel is $0.0010\% \times 100 = 0.10\%$!

This probability is far too restrictive in comparison to our initial desire: the Bonferroni correction applied here is too severe, and we are faced with the situation observed at the beginning of this section.

*The threshold given by the Bonferroni correction needs to be revised if the multiple tests performed are not independent.*

If we now consider only the 100 independent values, the adjusted Bonferroni correction is then: $10\%/100 = 0.10\%$ for each voxel. (We indeed find ourselves with the probability, once again, of obtaining one false positive voxel: $0.10\% \times 100 = 10\%$).

*The "corrected" Bonferroni procedure is obtained by dividing the chosen statistical threshold by the number of independent tests carried out.*

The "non-independence" of the different voxels in a functional image may result, for instance, from spatial smoothing, which is used as a pre-treatment of fMRI data. Smoothing entails averaging over the voxels, as in the (caricatural!) example presented above.

As a result, there will be a certain degree of spatial correlation in the image, which implies that the data in any given voxel will tend to resemble to data in the neighboring voxels.

Thus, there are fewer truly independent values than there are voxels.

When computing the Bonferroni correction, therefore, the methods most commonly employed in functional brain imaging take account of the correlation existing between the tests or, in other words, base the calculation on the number of "real" tests ("random field theory" is one such method).

## 4.5. Pre-processing and limitations of fMRI

*The statistical processing studied in section 4.4 can only be done if corrections are made to anatomical and functional images.*

*In this section, we list all of the artifacts that need to be corrected, and put forward solutions to remedy them.*

### 4.5.1. *Motion correction*

4.5.1.1. *Why is this correction necessary?*

It is inevitable that during an fMRI acquisition, the patient will move, because of the length of the exam. However, the increase in the signal expected at the moment of activation in relation to the rest state does not exceed 5%, as seen in section 4.2.1.3. Thus, any movement of the subject is highly damaging to a correct analysis of the results.

We can classify movements into two main categories.

### 4.5.1.1.1. Movements of physiological origin

Such movements cause a low-frequency noise to occur (respiratory movements: 0.25 Hz; cardiac movements: 1 Hz, etc.).

This noise is superposed on the hemodynamic response, which therefore becomes "less clear": hence, in the estimation $b_1$ of the weight $\beta_1$ of the regressor $X_1$, the variance of the residual error $\sigma^2$ will increase. In other words, the estimation of the model "loses accuracy".

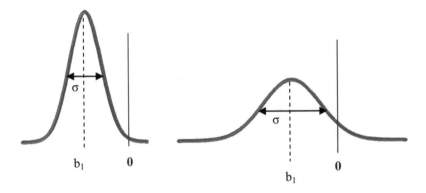

**Figure 4.43.** *Distribution of the weight $\beta_1$ around its estimator $b_1$ without (left) and with (right) noise*

Put differently, the loss of precision of the model implies that a "0" value for the regressors, indicating that they do not play a part in the signal (see Figure 4.45), becomes a possible value (i.e. one that is truly envisageable).

When detecting the pixels activated with a particular paradigm, the choice of a correct risk threshold (generally 5%), means that many pixels which have become "risky" are discounted: *there is a loss of sensitivity in the detection of brain activity.*

"*Mathematically speaking*", the choice of risk threshold determines the minimum error $t_\alpha$ for the pixel to be considered as active. In our case, the relative error $t_{obs}$, which is inversely proportional to the square root of the variance, becomes smaller. Fewer pixels will satisfy the condition $t_{obs} > t_\alpha$: therefore the number of activated pixels will drop.

4.5.1.1.2. Voluntary movements

The rather long duration of an fMRI paradigm makes it practically impossible to avoid movements of the subject. These movements may, therefore, give rise to false activation detections.

Let us now illustrate how this is possible.

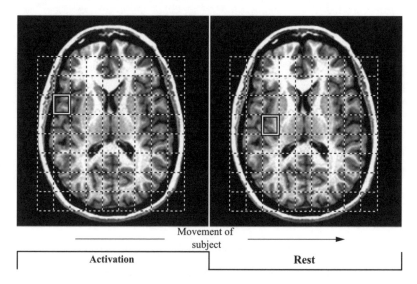

Figure 4.44. *Pixel examined and movement of the subject during the alternation period between activation and rest*

The variation of the signal in a given voxel may not be due to the change between activation and rest in the paradigm, but rather to the shifting of that voxel, e.g. from white to gray matter, when the subject moves.

If the shift of the voxel occurs seemingly in phase with the alternation imposed by the paradigm, statistical processing of the signal may conclude that the voxel is activated, with a large margin of probability! This voxel is therefore a "false positive" (which the statistical process regards as a "true positive"). *There is a loss of selectivity in the detection of brain activity.*

Thus, if the movements of the head are not properly corrected, the cerebral activity detected may ultimately turn out to be no more than a simple artifact of motion. The effects on a search for a cognitive zone are far from uncommon (see Figures 4.47 and 4.48).

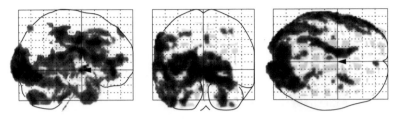

**Figure 4.45.** *Pixels detected as "activated" when the subject is moving*

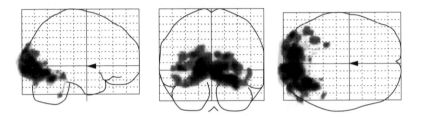

**Figure 4.46.** *Pixels detected as "activated" when the subject is not moving*

### 4.5.1.2. *Correction of artifacts of motion*

#### 4.5.1.2.1. Correction of artifacts caused by physiological movements

The artifacts due to physiological movements can be greatly lessened by *high-pass temporal filtering* (or by modeling the signal using regressors of no interest: see section 4.4.4.1).

#### 4.5.1.2.2. Correction of artifacts caused by voluntary movements

The procedure used to correct artifacts caused by the patient's voluntary movement is more complicated.

We need to *adjust all the images* ("intrasubject adjustment") to a reference image (often an average of several acquired images).

The most widely used method takes account of a movement of the head in the three spatial directions (three axes of translational motion) and three rotations around each of these axes.

We perform successive adjustments of the image using these different transformations (translations and rotations); we then evaluate the average $\varepsilon$ of the differences in intensity of the squared levels of gray between the adjusted image and the reference image: it is by minimization of that average that we obtain the definitive image.

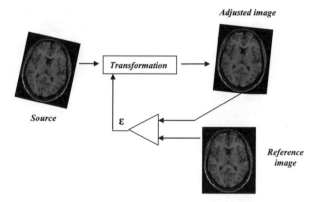

**Figure 4.47.** *Principle of adjustment of images when faced with voluntary movements*

Thus, we obtain a series of parameters indicating the translations and rotations necessary to adjust each image.

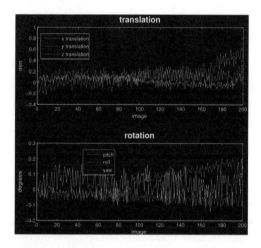

**Figure 4.48.** *Parameters of adjustment of each image in a translational and rotational orientation (taken from [DAV 09]). For a color version of the figure, see www.iste.co.uk/perrin/MRITech.zip*

NOTE.– The real procedure used for correcting motion artifacts is a little more complex. It involves creating new images: each new voxel will be estimated in the form of weighting of the voxels from the original image (weighting which will obviously result from the transformations estimated during the adjustment stage). This step is known as re-sampling.

Another possibility is to *model the movement in the form of regressors of no interest:*

– *Advantage:* this approach greatly decreases the number of false positives (increased selectivity).

– *Disadvantage:* if the movement is correlated with the experimental paradigm, certain genuine activations may not be detected (decreased sensitivity).

4.5.1.3. *Overview*

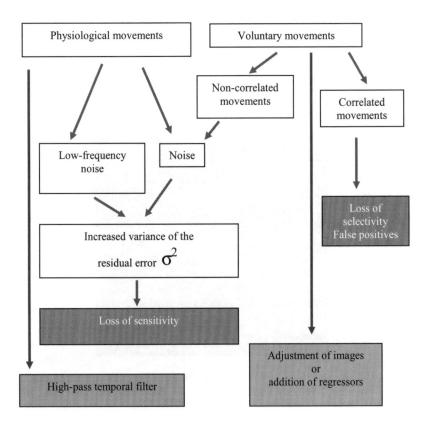

**Figure 4.49.** *Impacts and proposed solutions to correct the different types of movement in fMRI*

### 4.5.2. *Correction of delay in acquisition between slices*

By the acquisition of slices based on the EPI sequence, we are able to acquire the entirety of the targeted cerebral volume in a single TR (see section 4.2.3).

The set of slices making up that volume are acquired not simultaneously but successively: either in sequential mode (from the first to the last), or in interlaced mode (even-numbered slices followed by odd-numbered slices).

The signals measured in each voxel will therefore be temporally shifted depending on the slice to which they belong.

The following signals belong to the voxels deemed to be activated.

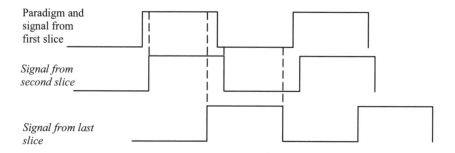

**Figure 4.50.** *Temporal positions of paradigm and of different signals obtained*

During the statistical processing of the data, we compare a simulation of the expected signal (with the regressors) to the signal from the voxel under study. If the temporal shift is not corrected, there will be a high number of false negatives (all the more so when the shift is great).

Thus, the shift observed for each voxel necessitates the definition of a model specific for each slice, with temporally shifted regressors.

In practice, this method is very infrequently employed. Practitioners tend to prefer to apply a *temporal adjustment* to the signal from each voxel, so that all of the signals appear to have been acquired simultaneously. (For this purpose, we perform a temporal interpolation which will not be detailed here).

### 4.5.3. *Correction of geometrical distortions*

Differences in magnetic susceptibility at the interface between two media (air/tissue or tissue/bone) can create a "parasitic gradient" $\beta_x$ or $\beta_y$ (in both directions of the plane).

The image is therefore expanded in both directions of a factor: $1+\dfrac{\beta_y}{\langle G_y\rangle}$ (direction of phase gradient) and $1+\dfrac{\beta_x}{\langle G_x\rangle}$ (direction of reading gradient).

$\langle G_x\rangle$ and $\langle G_y\rangle$ here represent the average pseudogradients or gradients.

In practice, only the direction of the phase coding gradient is affected by this artifact, because $\langle G_y\rangle$ is around 100 times smaller than $\langle G_x\rangle$.

Two solutions are therefore mainly used to solve these problems of distortion:

– *acquiring a phase map* to determine the parasitic gradient at each point and thereby find the corrections of position that need to be applied to the different voxels in the image (we shall not go into further detail about this technique);

– *performing a segmented acquisition.*

### 4.5.4. *Overall view of treatment in fMRI*

Below, we show all of the operations which may be carried out during an fMRI exam.

Certain treatments apply only to group study (rather than to individual examinations) which we shall discuss in section 4.6.

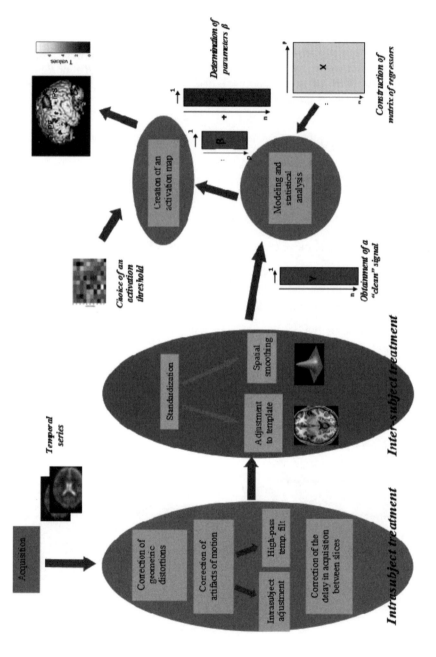

**Figure 4.51.** *Overall view of treatments to be performed during an fMRI examination*

## 4.6. Application of fMRI

*In order to comprehend the importance that fMRI has acquired in the study of cognitive phenomena, we need to focus both on its applications in the field of research (by way of group analysis) and its applications in clinical practice (diagnosis and treatment support).*

### 4.6.1. *Group analysis*

4.6.1.1. *Why a group analysis?*

An fMRI study (outside of clinical routine) is not generally limited to the study of a single subject, as has been the case with our investigations hitherto.

The objective is to move to the study of groups, which has the advantages of:

– enhancing statistical power. We can therefore compare the study of a group to that of a single subject, but using a greater number of measurements, and therefore a lower variance between the residuals $\sigma^2$ ;

– extrapolating the results obtained with one individual to the whole of a population.

Thus, we need to compare the activation maps obtained for n subjects and be capable of summarizing them in a group activation map.

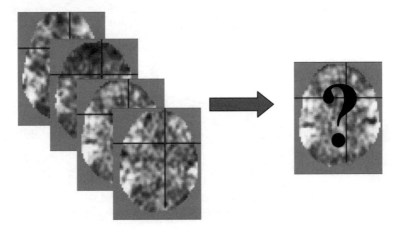

**Figure 4.52.** *Concatenation of results obtained on different subjects to create a map of group activation (taken from [KEL 08])*

4.6.1.2. *What problems may be encountered?*

We need to be able to define a cerebral morphology that is common to all brains.

Yet the major challenge posed by this group study is, quite precisely, the great morphological variability of the human brain!

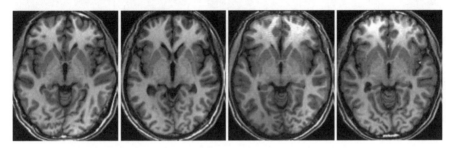

**Figure 4.53.** *Some transversal slices taken from different subjects (taken from [KEL 08])*

This step, which is necessary when performing group analysis, is known as "*standardization*".

It takes place in two stages:

– adjustment;

– spatial smoothing.

4.6.1.2.1. Adjustment

We wish to perform matching between the different brains so that the homologous functional regions are as closely matched as possible.

For this purpose, a reference image, the template, which represents an "average" brain, is used.

One of the best-known templates is the MNI (Montreal Neurological Institute) template.

During the operation of adjustment, various transformations can be used to adjust the shape of the brain under study to that of the template:

*Linear transformations*

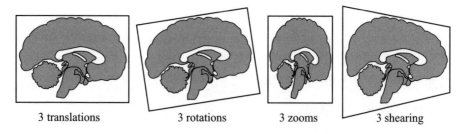

| 3 translations | 3 rotations | 3 zooms | 3 shearing |

**Figure 4.54.** *The different types of linear transformations used*

*Nonlinear transformations*

Nonlinear transformations are deformations to the whole of the brain whose field of displacement is more or less regular.

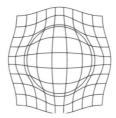

**Figure 4.55.** *Example of application of a nonlinear transformation on a grid*

*In summary*

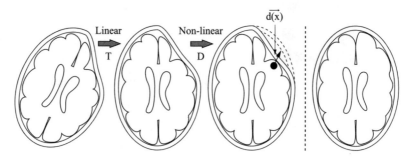

**Figure 4.56.** *Paradigm of transformations used in adjustment*

The best possible adjustment is obtained by seeking to minimize the quadratic mean of the differences in intensity between the anatomical image of the brain being studied and the template.

Thus, we can see that this operation of adjustment is identical to that which is used when correcting motion, except that additional transformations such as zooms, shears or indeed nonlinear transformations come into play.

4.6.1.2.2. Spatial smoothing

In order to correct the residual errors in adjustment, spatial smoothing is applied. This smoothing operation consists of *low-pass spatial filtering* by convolution of each image with an isotropic Gaussian filter whose width at the half-height is set by the user.

*Advantage*: the SNR is significantly improved

*Disadvantage*: loss of spatial resolution in functional images because smoothing mixes the signals from different voxels.

The user needs to choose the width of the Gaussian filter at the half-height so that it is seemingly identical to that of the activated zones being investigated.

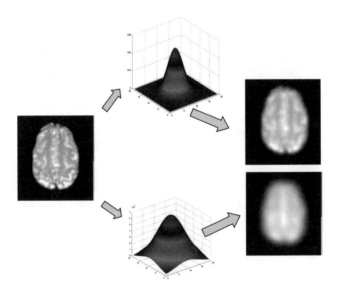

**Figure 4.57.** *Influence of the width of the Gaussian filter used for smoothing on the spatial resolution of the images*

When the size of the zones being sought is not known, we generally take a width at the half-height of two or three times the size of a voxel.

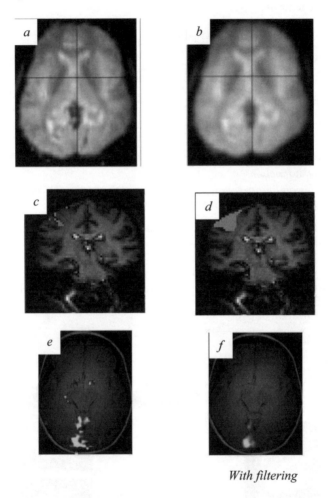

*With filtering*

**Figure 4.58.** *Anatomical and functional images without (a, c, e) and with smoothing (b,d,f) (taken from [MER 07]). For a color version of the figure, see www.iste.co.uk/perrin/MRITech.zip*

NOTE.– The decrease in the independence of the voxels due to spatial smoothing needs to be taken into account in the Bonferroni correction when making multiple comparisons.

### 4.6.2. *Disease*

#### 4.6.2.1. *Pre-surgical functional map*

The exeresis of tumors is an operation which is often subject to much discussion, because of the danger of damaging functional zones during the intervention.

Given this danger, fMRI can be a very useful ally:

– it facilitates the making of decisions regarding a possible intervention by localizing the functional zone on which the surgeon needs to operate, and giving indications about the extent of the tumor in relation to that zone;

– it facilitates intervention by way of neuronavigation techniques. *The pinpointing of the functional zones helps the surgeon to establish a plan as to how to approach the lesion and orientate him/her during the operation with respect to the "at-risk" zones.*

Two examples will be discussed here:

– pre-surgical motor function map;

– study of hemispheric dominance of linguistic skills.

#### 4.6.2.1.1. Localization of motor centers

Using simple motor tasks (flexion/extension of the fingers, toes and contraction of the lips), it is possible to pinpoint the primary motor cortex with the entirety of its organization.

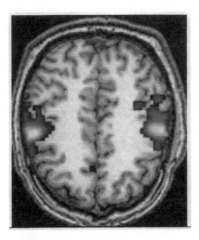

**Figure 4.59.** *Activations detected when moving the lips. For a color version of the figure, see www.iste.co.uk/perrin/MRITech.zip*

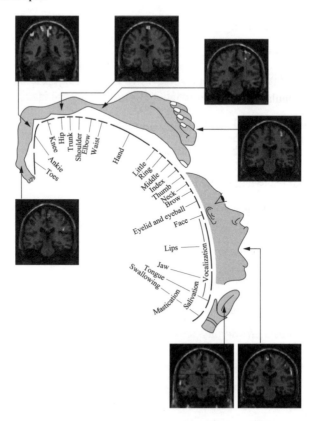

**Figure 4.60.** *Entirety of the motor cortex. The size of the different areas is modeled by the volume of each part of the body. For a color version of the figure, see www.iste.co.uk/perrin/MRITech.zip*

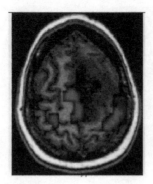

**Figure 4.61.** *Tumor situated on contact with the supplementary motor area (SMA). The SMA is used for the planning of complex movements (such as those involving the use of both hands, for instance). For a color version of the figure, see www.iste.co.uk/perrin/MRITech.zip*

4.6.2.1.2. Localization of linguistic areas

It is crucial to pinpoint the linguistic areas of the brain when a tumor is located:

– in the hemisphere dominant for language (see the next section);

– near to particular zones, such as:

- the inferior frontal gyrus: Broca's area (used for formulating a verbal response: this is the area where language is assembled and organized),

- the posterior third of the superior temporal gyrus: Wernicke's area (the area for reception, decoding and storage of language).

There is then a danger of permanent linguistic deficit as a result of surgical intervention.

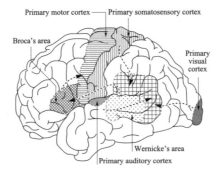

**Figure 4.62.** *Location of certain sensory areas*

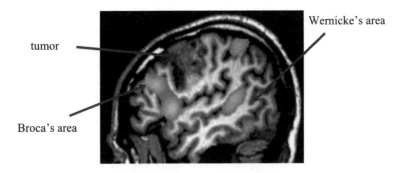

**Figure 4.63.** *Tumor and location of Broca's and Wernicke's areas.*
*For a color version of the figure, see www.iste.co.uk/perrin/MRITech.zip*

In the clinical case illustrated by Figure 4.62, the linguistic areas (here Broca's area and Wernicke's area) are a good distance away from the tumor. However, care needs to be taken with regard to the primary somatosensory cortex, because it is located just behind the tumor.

*How can we determine the hemisphere dominant for language?*

– There is a connection between manual laterality and hemispheric language dominance. Thus, the left hemisphere is dominant for language in 90% of right-handers and 70% of left-handers. These statistics, though, cannot provide any absolute certainty...

– The Wada test: up until only around a decade ago, the Wada test was the predominantly used technique. The procedure consists of selectively injecting a barbiturate into one carotid and then the other, and testing the contralateral hemisphere in the conscious subject. This is, of course, an invasive exam.

– One enormous advantage to fMRI is that it is non-invasive, and has been vindicated because it exhibits a strong correlation with the Wada test.

*Determination of hemispheric dominance of language using fMRI*

We determine all of the language-related areas by having the subject perform different tasks. When these different areas have been determined, we obtain the hemispheric dominance by calculating an index of laterality IL:

$$IL = \frac{n_{left} - n_{right}}{n_{left} + n_{right}}$$

with $n_{left}$ and $n_{right}$ being the respective numbers of voxels activated in the left hemisphere and in the right hemisphere.

For instance, if IL = +1: the patient presents exclusively left lateralization.

4.6.2.2. *Cerebral plasticity and functional recovery*

The brain is not a static system. Thus, a damaged brain (suffering from a vascular malformation, tumor, etc.) is able to modify its activity over more or less long periods of time. This capacity to adapt is called plasticity.

The total innocuousness of MRI means that the tests can be repeated over time, and therefore we can study cerebral plasticity over short periods, with regard to the motor system or the linguistic centers, for instance.

# Bibliography

[AUE 04] AUERBACH E.G., MARTIN E. T., "Magnetic resonance imaging of the peripheral vasculature", *American Heart Journal*, volume 148, issue 5, Elsevier, November 2004.

[BRE 03] BRETT M., PENNY W., KIEBEL S., "An Introduction to Random Field Theory", *Human Brain Function*, Ashburner J., Friston K., Penny W. (eds), Wellcome Trust Centre for Neuroimaging, London, 2003.

[CIU 13] CIUCIU P., The joint detection estimation framework, in BOLD functional MRI, UNATI meeting, 19th March 2013.

[DAV 09] DAVID O., "IRM fonctionnelle de petit animal: principe, acquisition, traitement", *JIRNFI Petit Animal,* Grenoble, 19 -23 October, 2009.

[GRA 07] GRAND S., LEFOURNIER V., KRAINIK A., BESSOU P., TROPRES I., CHABARDES S., HOFFMANN D., LE BAS J.-F., "Le point sur... - Imagerie de perfusion : principes et applications cliniques", *Journal de Radiologie,* Vol. 88, No.3-C2, p.444-471, March 2007.

[HIL 08] HILL C. S., Lambert C., "Random Field Theory", *Methods for Dummies*, Wellcome Trust Centre for Neuroimaging, UCL, London, October 2008.

[IRM] IRM EN IMAGE, available at: http://irmenimage.e-monsite.com/medias/album/images/47444046191-jpg.jpg.

[KEL 08] KELLER M., *Modélisation de l'incertitude spatiale pour l'inférence de groupe en IRM fonctionnelle,* http://www.agroparistech.fr/IMG/pdf/Keller.pdf., November, 2008.

[MER 07] MERIAUX S., Diagnostique d'homogénéité et inférence non-paramétrique pour l'analyse de groupe en imagerie par résonance magnétique fonctionnelle, Thesis, University Paris-Sud II, 2007.

[PER 09] PERLBARG V., Module "Imagerie cérébrale", December 2009.

[RAD 06] RADIOL J., Diffusion-weighted imaging of the brain: normal patterns, traps and artifacts, December 2006.

[ROY 11] ROYET J.-P., *Se souvenir des odeurs, ça se travaille*, Futura santé, http://www.futura-sciences.com/magazines/sante/infos/actu/d/medecine-souvenir-odeurs-ca-travaille-28630/, 11/03/2011.

[SAV 03] SAVATOVSKY J., MARRO B., Séquence de diffusion dans l'IRM cérébrale, http :www.chups.jussieu.fr/polys/radiologie/jrad/Diffusion.swf, Mars 2003.

[VLA 03] VLAARDINGERBROEK M.T., DEN BOER J. A., "Motion and flow", *Magnetic Resonance Imaging: Theory and Practice*, 3rd edition, Springer, Heidelberg, Germany, 2003.

[VOS 98] VOSSHENRICH R., KOPKA L., CASTILLO E.,  BÖTTCHER U., GRAESSNER J., GRABBE E., "Electrocariograph-triggered two-dimensional time-of-flight versus optimized contrast-enhanced three-dimensional MR angiography of the peripheral arteries", *Magnetic Resonance Imaging*, volume 16, issue 8, Elsevier, October 1998.

# Index

eigenvalues, 87-89, 92-95
eigenvectors, 87-89, 92
endogenous tracers, 111, 125-126
EPISTAR sequence, 127-128, 130
estimator, 154, 157-160, 170-174,
    183, 204

**F**

F test, 190
FACT method, 97
FAIR sequence, 127-128, 130
false positive, 198-200, 202-203,
    205, 208
fast flow, 7, 28, 36
filter, 150, 176, 182, 215
FLAIR sequence, 22, 71-73, 128
flow compensation gradient, 10-11,
    19, 22, 31, 47
free
    protons, 22, 121
    spins, 121-122, 130

**G**

gadolinium, 37- 40, 42-46, 105-108,
    110-111, 141
ghost image, 12-13, 18, 79-80
gradient echo (GE), 3, 149
group analysis, 198, 212-213

**H**

hematomas, 22, 28
hemispheric dominance, 217, 220
hemoglobin, 78, 104, 142
hypoperfusion, 110

**I**

inversion pulse, 71, 112-116, 119,
    121-131
ischemia, 63-65, 73-75, 78, 104,
    133-134

ischemic accident, 68, 70-71, 78-79,
    134-135
isotropic image, 64-65

**L**

labeling acquisition, 112-115, 117,
    121-123, 125-130
least squares method, 159
linear model, 166
low-frequency drifts, 184, 190

**M**

macrovascularization, 103, 133
magnetic susceptibility, 78, 105-106,
    111, 210
magnetization transfer, 120-128,
    130-131
microvascularization, 105, 131, 133
mismatch, 134, 137
morphological variability, 213
motor centers, 217
multi-slice, 7, 124, 127, 130, 149
myelin, 51

**N**

neurovascular coupling, 142-143
noise, 32, 98, 182-183, 185, 204
normal law, 156-157, 159-161, 164,
    201

**O,P**

occlusion, 43, 135, 137
paradoxical enhancement, 4-6, 12,
    21, 28, 46
penumbral zone, 77, 134
phenomenon of slice entry, 20
PICORE sequence, 129-130
potentials for action, 141